# Irrigation Management and Globalisation

# IRRIGATION MANAGEMENT AND GLOBALISATION

*By*
**Dr. M. Lakshmi Narasaiah**
M.A., Ph.D.
*Professor of Economics,*
*Co-ordinator, Department of M.B.A. and Commerce,*
*Special Officer,*
*Sri Krishnadevaraya University Post-graduate Centre,*
*Kurnool–518 002*
*Andhra Pradesh (India)*

DISCOVERY PUBLISHING HOUSE
NEW DELHI

First Published – 2006

Reprinted – 2025

ISBN: 978-81-8356-060-3



**Irrigation Management and Globalisation**

*Published by:*

**DISCOVERY PUBLISHING HOUSE**

4383/4B, Ansari Road, Darya Ganj
New Delhi-110 002 (India)
*Phone*: +91-11-23279245; 23253475; 43596065
*Mobile*: +91 9811179893 / +91 9871656464
*E-mail*: discoverybooksindia@gmail.com
orderdphbooks@gmail.com
namitwasan9@gmail.com
*web*: www.discoverypublishinggroup.com

*Printed at:*
Infinity Imaging Systems
Delhi

# Preface

Irrigated agriculture is up against an enormous challenge. India's population continues to grow at a tremendous rate. Over the next 10 years, there will be more extra mouths to feed. Water is becoming increasingly scarce for agriculture, with conflicting demands being made on limited supplies by the domestic and industrial sectors. The most attractive irrigation sites have already been exploited. Yet, irrigated agriculture will have to deliver average output increases of at least 3.5 per cent per year if future food demands are to be met in India.

Forty years ago in the 1950s and 1960s, India has worried about its capacity to produce sufficient food to feed its growing population. Episodes of food scarcity were not uncommon. It was dreaded that millions would die of hunger. Then dawned the miracle of the "green revolution". Cereal production increased by leaps and bounds, boosted by expanded irrigation, increased fertilizer application and modern crop varieties. The vigorous response in mobilizing financing towards boosting agricultural production paid off.

Given the importance of irrigation to the India's food supply and the vast resources already expanded on irrigation development, it is tragic that the actual performance of irrigation systems is so disappointingly low. In the post-green revolution era, it has become increasingly evident that the performance of irrigation systems, especially large-scale systems, is suboptimal whether measured in terms of achieving planned area targets or in terms of production potentials created by the physical works.

In many irrigation systems, the actual area irrigated is much less than the command area. Sharp inequities in water supplies between farmers in the head reaches of irrigation systems and those located downstream is another manifestation of poor performance. Investigations in the Tungabhadra Irrigation Scheme reveal that the tail-end of a major distributory commanding 25 per cent of the total area, received approximately 20-40 per cent of the targeted discharge while the upper reaches got more than their share.

Lack of maintenance has caused many systems to fall into disrepair, further inhibiting performance. Over time, distribution canals have become silted up, increasing the likelihood of breaching, damage to outlets and leading to salt build-up in the soil.

**Successful Farmer-managed Irrigation Systems**

Farmers have long demonstrated their potential capability to manage irrigation systems efficiently. Farmer-managed irrigation systems, (FMIS) also known as traditional, indigenous, communal or people' systems, are often classified "minor" or "small-scale" irrigation systems, although they may be found in command areas of 15,000-20,000 hectares.

**Dr. M. Lakshmi Narasaiah**

# Contents

1

# Irrigation Management
## *Facing the Challenge*

Irrigated agriculture is up against an enormous challenge. India's population continues to grow at a tremendous rate. Over the next 10 years, there will be more extra mouths to feed. Water is becoming increasingly scarce for agriculture, with conflicting demands being made on limited supplies by the domestic and industrial sectors. The most attractive irrigation sites have already been exploited. Yet, irrigated agriculture will have to deliver average output increases of at least 3.5 per cent per year if future food demands are to be met in India.

Forty years ago in the 1950s and 1960s, India was worried about its capacity to produce sufficient food to feed its growing population. Episodes of food scarcity were not uncommon. It was dreaded that millions would die of hunger. Then dawned the miracle of the "green revolution". Cereal production increased by leaps and bounds, boosted by expanded irrigation, increased fertilizer application and modern crop varieties. The vigorous response in mobilizing financing towards boosting agricultural production paid off.

Given the importance of irrigation to the India's food supply and the vast resources already expanded on irrigation development, it is tragic that the actual performance of irrigation systems is so disappointingly low. In the post-green revolution era, it has become increasingly evident that the performance of irrigation systems, especially large-scale systems, is suboptimal whether measured in terms of achieving planned area targets or in terms of production potentials created by the physical works.

In many irrigation systems, the actual area irrigated is much less than the command area. Sharp inequities in water supplies between farmers in the head reaches of irrigation systems and those located downstream is another manifestation of poor performance. Investigations in the Tungabhadra Irrigation Scheme reveal that the tail-end of a major distributory commanding 25 per cent of the total area, received approximately 20-40 per cent of the targeted discharge while the upper reaches got more than their share.

Lack of maintenance has caused many systems to fall into disrepair, further inhibiting performance. Over time, distribution canals have become silted up, increasing the likelihood of breaching, damage to outlets and leading to salt build-up in the soil.

**Successful Farmer-managed Irrigation Systems**

Farmers have long demonstrated their potential capability to manage irrigation systems efficiently. Farmer-managed irrigation systems, (FMIS) also known as traditional, indigenous, communal or people' systems, are often classified "minor" or "small-scale" irrigation systems, although they may be found in command areas of 15,000-20,000 hectares.

Many successful farmer-managed irrigated systems which have been functioning effectively for hundreds of years, represent a valuable, accumulated investment. They are also reservoirs of largely untapped irrigation management experience.

Research has revealed that FMIS contribute to the production of a significant portion of the subsistence food supply. Ground-water irrigation systems, often found in areas affected by drought, play a strategic role in promoting food security. In India ground-water development which is increasing in importance is predominantly farmer-managed. In this country an estimated 20 million hectares are already under ground-water irrigation. When mismanaged though, ground-water irrigation could impact negatively on the environment.

In addition to the hectarage under farmer-managed ground-water irrigation, farmer-managed tank irrigation systems cover about 8.5 million hectares in this country. Farmer-managed irrigation systems have also allowed intensification of agriculture to partially meet the food needs of rapidly growing populations.

**Little Awareness**

Yet, despite the widespread interest in irrigation management turnover throughout India there is very little documentation about the processes used and the results obtained from irrigation management turnover. Many policy-makers do not know how to turn over management of their irrigation institutions in an effective way. They typically have very little, if any, awareness of the range of organisational options which may be suitable under different conditions. There is an urgent need, therefore, for a systematic, comparative assessment of the range of approaches being used, constraints to implementation and the impacts on performance of transferring management to non-governmental or farmers' organisations.

Successful irrigation in the future will be that which supports much higher levels of agricultural productivity, enhances responsiveness to more diversified and dynamic crop markets, stimulates more profitable irrigated agriculture for wide numbers of rural poor, substantially improves water use efficiency and supports and sustainable use of scarce land, biomass and water resources.

In the coming years the irrigation sector will be in ferment, with decision makers, agency managers and farmers needing better information and strategic processes to make intelligent choices in the management of irrigated agriculture.

2

# Strategies for Improved Water Management

Since all surface and sub-surface water related activities are closely interlinked through upstream-downstream relations in river-basins and aquifers, the whole river-basin must be considered in all water development policies and research.

River-basin authorities should be established, for both national and international basins, and should have regulatory powers over iner-sectoral allocation of water, enforcement of water quality standards, arbitration in disputes, and compensation procedures, International basin authorities should be composed partly of representatives of national basin authorities.

Water policies should encompass underground water, surface water, and direct utilisation of rainfall, as an integrated continuum requiring common strategies.

**Participation of Users and Stakeholders**

Governments should strengthen stakeholder participation in the allocation and management of water resources by means of:

1. Identifying and establishing water rights for all users, following economic, cultural, environmental and social considerations;
2. establishing compensation procedure (both national and international);
3. using impact assessment as a tool to identify livelihood rights and ecological considerations.

Stake-holder participation in water reallocation and management should be used to reconcile multiple or contradictory objectives and to reduce conflicts over ownership or control of water.

**Environmental Protection**

To guide development actions and to monitor changes, basin-wide mapping of "hot spots" (areas of special problems or vulnerability with respect to water quality) should be undertaken; sources of major pollutants should be identified and published. Water bodies which have special ecological or societal significance should be preserved.

Training of professionals for the water sector should develop new environmental dimensions, and should address such issues as; conservation of water; responses to water scarcity; disaster preparedness.

**Research into the impacts of Water Policy**

The impacts of water policies, and policy changes, are not well understood. There is a need for structured research on impacts rather than reliance on anecdotal evidence.

Water development actions can have negative effects on equity, and can injure individuals or groups even while bringing wider benefits. There is a need for analysis of responses and solutions to these effects.

Externally defined systems of water rights will lack legitimacy and will fail to meet the needs of all users, if they do not take account of users' perceptions and customs. National Research Institutes should therefore develop and apply methodologies for assessing users perceptions and customary rights.

**Capacity-Building for Integrated Management**

The development of river-basin authorities will require the training of their personnel in: coordination; planning of systems for monitoring and evaluation; sensitisation to new responsibilities; integrated land and water resources uses; use of modelling tools; implementation of guidelines; procedures for stake-holder participation.

Politicians and polity makers need to become more aware of issues concerning; water scarcity and water needs; equity of access to water; water rights; sustainability; service improvement; enhancing the performance of water systems; stakeholder participation.

**External Assistance for Development**

Partnerships or "twinning" arrangements, should be promoted between new river-basin authorities and existing, successful ones.

External donor assistance programmes should focus on users' groups, especially among the disadvantaged and the poor. They should take a basin perspective, recognizing existing institutional arrangements and inter-sectoral requirements. Within this frame work they should foster and encourage mechanisms for devolution of responsibilities, tasks, and authority to local level users' groups.

External help is needed to develop and evaluate water data-bases, agricultural knowledge systems, methodologies and processes for water development and evaluation, and other information areas.

**Emergencies and Conflict Reduction**

National governments and river-basin authorities should develop contingency plans for the management of their responses to natural disasters, prolonged droughts, and other emergency situations in which normal procedures of water allocation and management may require temporary modifications.

Policies and actions in regard to shared water sources should be improved, with emphasis on consultation, coordination and sharing of information.

3

# Population Growth and Cropland

Since mid-century, global population has grown much faster than the cropland area. The trend is likely to continue in the next century, dropping cropland per person to historically low levels. The ever smaller per capita cropland bases will make food self-sufficiency impossible for many countries, and will test the capacity of international markets to meet a growing demand for imported food.

For millennia, farmers satisfied rising food demand by bringing new land under the plow. But by mid-century cropland expansion could no longer meet the food needs of an increasingly populous and prosperous world. The 10,000 year era of steady expansion was over, and a new era began that stressed raising land productivity. As this high-yielding era shows signs of faltering, concern over the shrinking supply of cropland per person looms ever larger.

Since mid-century, grain area—which serves as a proxy for cropland in general—has increased by some 19 per cent, but global population has grown 132 per cent, seven times faster. Largely as a result, grain area per person has fallen by half since 1950, from 0.24 to 0.12 hectares. Assuming that grain area remains constant, grain area per person will fall to 0.07 hectares by 2050. In crowded industrial countries such as Japan, Taiwan, and South Korea, grain area per capita today is smaller than the area of a tennis court.

As grain area per person falls, more and more nations risk losing the capacity to feed themselves. Having already seen per capita grain area shrink by 40-50 per cent between 1960 and 1998, Pakistan, Nigeria, Ethiopia, and Iran can expect a further 60-70 per cent loss by 2050—a conservative projection that assumes no further losses of agricultural land. The result will be four countries with a combined population of more than 1 billion whose grain area per person will be only 300-600 square metres, less than a quarter of the area in 1950.

The historical record suggests that such a small area per person will send a substantial share of a country's people to world markets for their food. Consider the experience of six countries in East Asia whose per capita grain area currently ranges from 200 to 600 square metres per person. Sri Lanka relies on imports for more than a third of its grain, while Japan, Thiwan, South Korea, and Malaysia buy more than 70 per cent of their grain from abroad. North Koera is the only one of the six that does not import heavily (it gets less than 20 per cent of its grain requirements from abroad), but its population is poorly fed-indeed, on the verge of starvation.

The concern is that population growth will push many nations—not just the four fastest-growing ones—below the 600-square metre-threshold in coming decades. In Asia alone, where grain area per person stands at 800 square metres, 16 countries are poised to cross this threshold by 2050, and many of them much sooner. As this process unfolds, the number of people who will turn to foreign markets for their food will likely jump sharply. These countries will find an increasingly tight international grain market, with nations from the Middle East, North Africa, and other regions already buying a third or more of their grain overseas.

In addition to per capita losses, population growth can lead to degradation of cropland, reducing its productivity or even eliminating it from production. As a country's population density increases and good farmland becomes scarce, poor farmers are forced onto ecologically vulnerable land such as hillsides and tropical forest. In the Philippines, for example, hillside agriculture accounted for only 10 per cent of all agricultural land in 1960, but 30 per cent in 1987. Because it is highly erodible, hillside land is easily damaged; worldwide, some 160 million hectares of hillside farmland—11 per cent of cropland—were characterised in 1989 as "severely eroded." Similarly, population pressure can force peasants to overfarm the poor soils of tropical forests. After being cleared and farmed for a few years, these soils typically require fallow period of 20-25 years, but population pressures keep poor farmers on the same land for far longer than the soils can support, cutting fallow period to just a few years in some areas of tropical Africa and Asia.

Finally, population pressures on a fixed base of land can result in rural landlessness. In Bangladesh, for example, landlessness among rural households rose from 35 per cent in 1960 to 53 per cent in the early 1990s. Interestingly, Bangladesh is regarded as a success in slowing population expansion, as its growth rate declined from 2.8 per cent in the late 1970s to 1.5 per cent in the early 1990s. But its success came too late to prevent the increase in rural landlessness, highlighting the need to work sooner, rather than later, for population stabilisation.

# 4

## Solving Conflicts Over Water Uses

There are few issues that have greater impact on the life of mankind and the planet as a whole than the management of our most important natural resource water. This has only been realised in detail more recently by the general public as well as by many planners and decision-makers. Around the world they have begun to appreciate the critical importance of a reliable water supply for their future survival and sustainable development. Rivers are lifelines in countries like India serving different uses such as transport, agriculture, fisheries, personal hygiene and others. Conflicts over use of water resources must be settled through better water management policies.

In many localities of the Earth water-related problems have become extremely acute, even critical. In some places they are the source of social instability and are a threat to international security. There is not the slightest doubt that, with further population increase and under the "water business-as-usual" scenario, these problems will become ever major acute, thus creating ever more instability. After decades of water waste, water pollution, and inability to provide basic water services to the poor we must fundamentally change the way we think about and manage water. We have to realise that water can no longer be considered to be a cheap and plentiful resource, which can be used abused or squandered without much concern for further human welfare.

As fresh water is becoming scarce, that is when there is not enough water to satisfy all demands, competition develops.

Besides the well known tensions at international level over limited water supplies there is an-increasing and in many cases a far more important competition over water arising within countries, between sectors. Such competition, for instance, occurs between farmers for irrigation water and between farmers and non-agricultural users of water such as cities (including industries and power plants) and environmental concerns (recreation, fish and other wildlife). Because of this increasing competition irrigated agriculture around the world, but especially in developing countries, faces important challenges in the coming decades, on the one hand, it has to provide a major share of the required increases in food and fiber production to meet the objectives of poverty alleviation and development. On the other hand. It is threatened by water shortages arising out of increasing competition from domestic, industrial and other sectors. This situation is further worsened by dwindling financial resources available for capital expenditure as the cost of new irrigation schemes increase.

**Land Under Irrigation**

On a regional basis, it is estimated that for example around 60 per cent of the value of crop production in Asia is grown on irrigated land. The irrigated sector performs an essential task in meeting the basic food needs of billions of people. It has provided more than half of the two most important basic staples and close to a third of all food crops. In the future, the irrigated sector will have to provide an even larger proportion of the total food output. The questions arises whether irrigated agriculture will be in the position to provide the extra food needed to feed a growing population despite an increasing water scarcity and inter sectoral competition. There is no general, no easy answer to this question. But the following implications are foreseeable.

The irrigation sector has to recognise that economic structures are dynamic and not static. The economics of many countries have undergone considerable change in the last decades. In connection with this change agriculture is losing its leading role in economic development and this is, besides other things, affecting the allocation of water resources between agriculture and other users. The same applies to many other countries around the world, especially developing countries in arid and semi-arid climates.

**New Water Policies are Needed**

Because of the growing competition for ever-scarcer water resources governments, and water authorities are forced to change water policies. Objectives of the new policies are demand-decreasing and demand-shifting. Irrigated agriculture, by far the dominant water user, will be strongly effected by such policy changes.

Water has an economic value in all its competing uses and should be recognised as an economic good. Agriculture will in future not any more get water free of charge as it did in the past. Farmers will have to pay for the water, and costs will be increasing steadily over the years to come.

There will be a reduction of the role of governments in rural water projects and an increasing importance of local user groups. Experience shows that an important solution to water related problems is to give users the responsibility for developing and managing the water resource. This requires that farmers become properly organised in water user associations able to discharge their new responsibilities, and that they have security of tenure to the land they farm and irrigate.

During the recent decade growth in crop productivity in irrigated areas has slowed, and competition for water for non-agricultural uses has increased. These developments place strong demands to develop water resource policies to maintain growth in irrigated agricultural production:

- facilitate efficient allocation of water across sectors and final demands; and
- reverse the ongoing degradation of the water resource base, including the watershed irrigated land base, and water quality.

What water policies can lead to efficient increases in irrigated production while reducing resource degradation in the irrigated areas in developing countries and releasing water for growing non-agricultural demands? What policies can be implemented to conserve water in non-agricultural uses to reduce competition between sectors?

**Questions to be Answered**

Among other the following questions still have to be answered:

1. What are the implications of growing competition between agricultural and non-agricultural uses of water for the availability and productivity of water in agriculture?

2. How to use regulations, water prices, pollution taxes and effluent charges to encourage water conservation and pollution control in industries and households?

3. What are the production, equity, employment generation and income impacts of alternative water allocation mechanisms in different agroeconomic and scarcity environments? What investment and administrative costs are associated with different mechanisms? What is the impact of resource allocation methods on water use, cropping patterns, crop yields, and fertilizer and other input use, capital investments, farm income, and environmental degradation?

4. What is the connection between alternative allocative mechanisms and the environmental externalities caused by irrigation, including water logging, salinisation, ground water recharge and ground water mining?

5. How will food security in low income countries be effected by changing water policies?

In the context of the growing competition for water it is important to enable policy makers to evolve a policy-mix which will result in efficient, equitable and environmentally sustainable allocation of water resources to user sectors. To achieve this, extensive research work is needed.

5

# Water Facts and Findings on Large Dams?

## Water Facts

Today, around 3800 $km^3$ of fresh water is withdrawn annually from the world's lakes, rivers and aquifers. This is twice the volume extracted 50 years ago. World population has passed 6 billion. Projections say that it will reach a peak of between 7.3 billion and 10.7 billion around 2050 before total population begins to stabilise or fall.

50 litres per person per day (or just over 18.25 $m^3$ a year) covers basic human water requirements for drinking, sanitation, bathing and food preparation. In 1990, over a billion people had access to less than 50 litres of water a day. Agriculture accounts for about 67% of withdrawals, industry users 19% and municipal and domestic uses account for 9%.

One third of the countries in water stressed regions of the world are expected to face severe water shortages this century. By 2025 there will be approximately 6.5 times as many people—a total of 3.5 billion living in water-stressed countries. By the end of the 20th century, there were over 45,000 large dams in over 150 countries. The average large dam today is about 35 years old. Since average construction periods generally range from 5 to 10 years, this indicates a worldwide annual average of some 160 to 320 new large dams per year.

During the 1990s, an estimated $32-46 billion was spent annually on large dams, four-fifths of it in developing countries. Of the $22-31 billion invested in dams each year in developing countries, about four fifths was financed directly by the public sector. About one-fifth of the world's agricultural land is irrigated, and irrigated agriculture accounts for about 40% of the world's agricultural production.

Half the world's large dams were built exclusively or primarily for irrigation, and an estimated 30 to 40% of the 271 million hectares of irrigated lands worldwide rely on dams. Dams are estimated to contribute to 12-16% of world food production. Hydropower currently provides 19% of the world's total electricity supply, and is used in over 150 countries with 24 of these countries depending on it for 90% of their supply.

Floods affected the lives on average, of 65 million people between 1972 and 1996, more than any other type of disaster, including war, drought and famine. There or 261 water sheets that cross the political boundaries two are more countries. A number of key international rivers lack of basin-wide agreement that defines a process for establishing equitable water use between riparian States.

**Cost Effectiveness**

Cost performance data confirms that large dam projects often incur substantial capital cost overruns. The average overrun was half again as much as the projected cost. The bulk of hydropower projects have delivered power within a close range of pre-project targets but with an overall tendency to fall short of targets.

At current rates, water fees are rarely sufficient to recover both capital and recurrent costs for water supply systems in many developing countries. Growing concern over the cost and effectiveness of large dams and related structural measures as long term responses to floods has led to support for integrated flood management as opposed to flood control.

Multi-purpose schemes are inherently more complex and many experience operational conflicts that contribute to under-performance on financial and economic targets. Substantive evaluations of project performance are few in number, narrow in scope, and poorly integrated across impact categories and scales.

**Ecological Costs**

Dams, inter basin transfers, and water withdrawals for irrigation have fragmented 60% of the world's rivers. As a physical barrier the dam disrupts the movement in upstream and downstream species composition and even species loss. In Africa, the changed hydrological regime of rivers has adversely affected floodplain agriculture, fisheries, pasture and forests that constituted the organising element of community livelihood and culture.

Problems may be magnified as more large dams are added to a river system, resulting in an increased and cumulative loss of natural resources, habitat quality, environmental sustainability and ecosystem integrity. Good site selection, such as not building large dams on the main-stem of a river system, and better dam design also played significant roles in avoiding or minimising impacts. The economic appraisal techniques such as risk and distributional analysis were still mandated for only 20% of large dam projects even in the 1990s.

**Social Costs**

At the planning and design stage, an important social impact is the delay between the decision to build a dam and the onset of construction. This can result in communities living for decades starved of development and welfare investments.

The overall global level of physical displacement could range from 40 to 80 million. In India and China together, large dams could have displaced between 26-58 million people between 1950 and 1990. Little or no meaningful participation of affected people in the planning and implementation of dam project—including resettlement and rehabilitation has taken place.

Empowering people, particularly the economically and socially marginalised by respecting their rights and ensuring that resettlement with development becomes a process governed by negotiated agreements is critical to positive resettlement and rehabilitation.

Poor accounting in economic terms for the social and environmental costs and benefits of large dams implies that the true economic efficiency and profitability of these schemes remains largely unknown.

The direct adverse impacts of dams have fallen disproportionately on rural dwellers, subsistence farmers, indigenous people, ethnic minorities, and women. Where costs and benefits accrue to different groups, the standard procedures for adding up and discounting the expected costs and benefits do not provide an appropriate measure of changes in social welfare.

**Financing**

Almost 2 billion people, both urban and rural poor, have no access to electricity at all. Most efficiency measures and technologies are cost-effective at today's electricity prices and the use of full environmental and social costing of electricity supply options makes them even more so. Among advanced technologies in research and development, micro turbines and fuel cell show the greatest near and mid-term promise.

Total financing for large dams from multi-lateral and bilateral development banks comes to more than $4 billion annually at the peak of lending during 1975-84. Although the proportion of investment in dams directly financed by bilaterals and multilateral was perhaps less than 15%. The total investment in dams by the multilaterals and bilaterals since 1950 is approximately $ 125 billion.

6

# A Breakthrough in the Evolution of Large Dams?

### Back to the Negotiating Table

"The problem is not the dams. It is the hunger. It is the thirst. It is the darkness in a township". With these plain words, former South African President Nelson Mandela summed up the World Commission on Dams (WCD's) motives in a speech at the presentation praising its work. His position is similar to that of the many other representatives of the South who were present: he demands a right to development. But Ms Medha Patkar, of India, a WCD Commission member and founder of the Struggle to Save the Narmada River (Narmada Bachao Andolan) anti-dam movement, takes a contrary view. "The problems of the dams are only a symptom of the larger failure of the unjust and destructive dominant development model," she says. We need to challenge "the forces that lead to the marginalisation of a majority through the imposition of unjust technologies like large dams".

### Stagnation of Dam Building

The WCD is a unique experiment in reaching consensus. It began in April 1997 when with the support of the World Bank and the World Conversation Union (IUCN), 39 representatives of diverse interests met at a workshop in Gland, Switzerland. At this point the various positions of the participants from governments, the private sector, international financial institutions, civil society organisations and affected people were cast in stone. The Manibeli

Declaration in June 1994 of 326 activist groups from 44 countries had called for an immediate moratorium on World Bank funded large dams until a comprehensive, independent review of all Bank funded projects had been conducted. International financial institutions were actually no longer able to fund further large dams in the face of public criticism. Enervated by steadily growing protests despite continual tightening of social and environmental standards, the institutions' representatives wearily likened the dispute to a football match in which somebody kept moving the goalposts. But one proposal to emerge from the meeting in Gland was for all parties to work together in establishing the World Commission on Dams.

The WCD began its work in May 1998 under the chairmanship of Prof. Kader Asmal, then South Africa's Minister of Water Affairs and Forestry. Its 12 members were chosen to reflect regional diversity, expertise and stakeholder perspectives. But the Commission ran the risk of failure right from the start due to their confrontational attitudes. The members, who spent 2 years 6 months jointly organising hearings, consultations and case studies and analysing more than 100 existing large dams, could not be more disparate.

**The Current Situation**

A large dam is a dam with the height of 15 metres or more from the foundation. If dams are 5-15 metres high and have a reservoir volume of more than three million cubic metres, they are also classified as large dams. Using this definition, there are more than 45,000 large dams around the world, almost half of them in China. They were built in the 20th century to meet the constantly growing demand for water and electricity. On a global scale, hydropower dams account for about 20 per cent of electricity generated, and in 24 countries, including Brazil, Democratic Republic of Congo, Zambia and Norway, hydropower covers more than 90 per cent of national electricity supply needs. Half the world's large dams were built solely or mainly for irrigation. Between 12 per cent and 16 per cent of world food production is based on dams, and as reservoirs they provide protection against floods.

Unfortunately, this impressive balance is counteracted by comparably significant problems. Construction of large dams is a major intervention in the ecosystem of rivers and the lives of many people. The WCD estimates that some 40-80 million people, mostly indigenous peoples, have been displaced by reservoirs worldwide and robbed of their livelihoods from fishing or farming. Serious conflicts are simmering between neighbouring countries because dams have turned off the water supply for downstream states.

The late Indian Prime Minister Jawarlal Nehru once said: "Dams are India's new temples." Right up to the 1970s, large dams were seen as the synonym for development and economic progress. Dam-building reached its peak between 1970 and 1980, when an average of two to three new large dams per day were commissioned. But a considerable number of the dams analysed by the WCD have fallen short of their technical and economic objectives. Construction cost overruns averaged 56 per cent. Many dams have had negative ecological impacts, and the disadvantages for people living downstream were mostly not taken into account. The planning of dams did not examine sufficiently possible alternatives for meeting power and water needs. There were hardly any retrospective evaluations of dam projects.

**A New Framework for Decision-Making?**

Despite this sobering stocktaking, the WCD arrives at an astonishingly simple finding: dams are primarily a means to an end. Their task is to improve the well-being of the people on a sustainable basis. This improvement should be economically acceptable, socially just and environmentally sound. If this goal can be achieved by a dam, its construction should be supported. Where alternative options offer a better solution they should be the preferred choice.

The Commission based its work on a set of five crore values for future decision-making: equity, efficiency, participatory decision-making, sustainability, and accountability. With regard to legal aspects and the extent of the potential risks for those involved, the WCD proposes development of an approach based on recognising rights and assessing risks. All risk-bearers should have a place at the negotiating table.

The WCD also recommends seven strategic priorities for decision-making: gaining public acceptance; comprehensive options assessment; reviewing existing dams; sustaining rivers and livelihoods; recognizing entitlements and sharing benefits; ensuring compliance; and sharing rivers for peace, development and security. These priorities are reinforced by 23 practical criteria and guidelines which can be adopted, adapted and applied by all actors involved in the dam controversy. For instance, the WCD suggests analysing points at issue together with the people affected by existing dams and developing joint proposals for solutions. People affected by new dam projects should be among their favoured beneficiaries, and their claims should be made legally binding.

The report offers a comprehensive compilation of knowledge, which previously was limited to individual case studies or a narrow specialist context, on the social, economic, technological and ecological problems and impacts of large dams. That makes the report a central reference which helps greatly in bringing objectivity into the debate. Provision of an analytical framework and strategic options is certainly an important step. But with regard to its task of developing internationally valid criteria and guidelines for the planning, design, appraisal, building operation, monitoring and shutdown of dams, the report remains very general. What will be decisive here will be to practice with the relevant actors the method and content of the suggested mediation process on the basis of specific cases.

**Deeds Must Follow Words**

The WCD's work must now be made useable for the private sector, civil society and development purpose. What is required is a discussion process involving all major actors.

The objective of this process must be the development of practical and effective guidelines in addition to the current standards.

The WCD calls on bilateral development organisations and multilateral development banks to support only dam projects that have resulted from an open process of examining various options. The parties should observe the WCD guidelines. Measures to save water and power should be examined and, if applicable, be promoted.

Private sector companies should publish guidelines on corporate behaviour and acknowledge the WCD principles, criteria and guidelines. Further, the private sector should draw up and implement voluntary codes of conduct, management systems and certification procedures, such as the internationally recognized standard for environmental management (ISO 14001). The OECD's Anti-Corruption Agreement should be observed, and declarations of honesty incorporated in contracts. Business associations should develop processes to monitor compliance with the WCD guidelines.

NGOs should primarily check compliance with agreements and assist aggrieved parties to seek compensation. They should also assist in identifying relevant stakeholders for dam projects, using the rights and risks approach. Finally the NGOs should build up support networks and partnerships between them.

**Importance of Information**

But do these noble proposals provide the whole answer? the WCD process is based on the opportunities for personal development of every individual in an open society. But it is not enough to build on the negotiating abilities of the potentially affected alone because only specialists can anticipate the complex impacts of dams. Therefore participation presupposes that the mediation process contains a substantial informative component.

Relying solely on a mediation process is also not sufficient in providing for social impacts. There is no generally recognized method to determine the value of 'goods' in a subsistence economy. Without objective criteria, the moral demands on both sides are extremely high. Strategically-motivated behaviour will then not be prevented even if all participants agree readily that the subjective standard of living of people affected by a dam should be improved or at least maintained.

Meditation processes make sense only if agreements are observed. According to the WCD analyses, lack of compliance with agreements is the main cause of the negative social impacts of dams. In the case of dam projects co-financed by international donors, disbursements of funding instalments could depend upon independent evaluations that confirmed compliance. Ensuring compliance is much more difficult in the case of projects financed by the private sector. Because there is no independent institution that could assume the role of arbitrator, it must be in the business world's own interests to act responsibly in ecological and social terms. To be credible, it must provide transparency and independent certifiers.

Due to the opposing interests involved, no-one should expect reaching consensus to be easy. But the example set by the WCD is not the only reason for hope. Taking a closer look at it, the model offers significant advantages for all participants. Partner countries and development organisations wish to continue to use the potential for development which dams will also offer in the future. The private sector will also continue to build and operate dams, and for that they need planning certainly. The advantages for NGOs and the people affected are also obvious.

7

# End of Controversy on Large Dams?

Was it worth the effort? The energy, the time, the money invested? Quite a number of people probably put this question to themselves on the 16th of November 2000 when Nelson Mandela launched the final report of the World Commission on Dams (WCD) in London. In an extraordinary process that lasted two and a half years, dam proponents and opponents worked intensely together. Twelve commissioners had tried hard to come up with a consensus on the effects of large dams in the past, and with recommendations for future sustainable planning on water and energy issues. And surprisingly enough: the Commission succeeded. The cost of ten million dollars was financed by governments, international agencies, the private sector, NGOs and various foundations—this is also a novelty. Dam-affected people, non-governmental organisations, companies, consultants, politicians etc. contributed to the processes with their know-how and by giving support and additional resources to the Commission's work.

The establishment of the multi-stakeholder commission was the result of a growing and aggravating controversy about the social, ecological and often enough also economic costs of large dams.

**Global Review Shows Faulty Planning Processes**

The WCD Global Review of large dams proves that there were good reasons for massive resistance against large dams in the past. As the report says: "In too many cases an unacceptable price has been paid... especially in social and environmental terms, by people displaced, by communities downstream, by taxpayers and by the natural environment" While funders of large dams, political institutions, consultants and companies claimed in recent years that they had learned from their faults and improved their performance, the WCD highlights that, while policies and assessment procedures have been improved, it appears that business-as-usual too often continued to prevail:

- Even in the 1990s, impacts of downstream livelihoods were not adequately assessed or accounted for in the planning and design of large dams.
- Participation and transparency in planning processes for large dams was neither inclusive nor open, and while actual change in practice remains slow, even in the 1990s, there is increasing recognition of the importance of inclusive processes.
- Where opportunities for the participation of affected people and the undertaking of environmental and social impact assessment have been provided they often occur late in the process, are limited in scope and even in the 1990s their influence in project selection remains marginal.

**Dams Hindered Human Development**

The WCD states that "dams have made an important and significant contribution to human development, and the benefits derived from them have been considerable". A viewpoint that dam-affected people and NGOs hardly share, having in mind the experiences of the past. Medha Patkar, WCD Commissioner and activist in the struggle to save the Narmada river in India wrote in her comment to the Report: "Within the value framework the Commission propagates—equity, sustainability, transparency, accountability, participatory decision-making and efficiency—large dams have not helped attain, but rather hindered 'human development'.

The WCD also puts an end to the old viewpoint that the violation of human rights and the social costs that some have to pay can be justified by the benefits for the others. The idea that society as a whole would profit from so-called development that were benefiting from 'trickle down effect'. In fact this development model tends to aggravate social inequities and encourage environmental destruction leaving the rich better of, but the poor more marginalised and resentful. It is not a sustainable model.

**It's Not Just About Dams**

Although it is called the World Commission on Dams the issues before the Commission are much broader. The discussion about large dams is inevitably linked with the need to find solutions for supplying water and energy in the future. To do this in a sustainable way is one of the challenges of our time.

The WCD defined 5 core values that are based on internationally accepted norms like the Universal Declaration of Human Right to Development and the Rio Declaration on Environment and Development:

- Equity
- Efficiency
- Participatory decision-making and Accountability

These values are not really new and some of them have been agreed upon decades ago, but the Global Review of the WCD showed that 'in real life' we are still far from implementing them.

The rights and risks approach of the WCD must be mentioned as an important tool: while funders and dam-builders talk a lot about their (mostly) financial risks, risks of the dam-affected people were not a big issue in the past. The Commission makes an important distinction: while the former take a voluntary risk and have the possibility to decide on whether or not they want to take it, the latter take involuntary risks and so far had hardly any option in deciding on whether or not they are willing to incur them. Rights must not be violated in order to serve the needs of other people. Respecting human rights is the minimum basis for discussion and is not negotiable.

To get better results in the future the WCD defined seven strategic priorities for decision-making:

- Gaining Public Acceptance
- Comprehensive Options Assessment
- Addressing Existing Dams
- Sustaining Rivers and Livelihoods
- Recognising Entitlements and Sharing Benefits
- Ensuring Compliance
- Sharing Rivers for Peace, Development and Security

According to the WCD there are several fundamental strategic points in the decision-making process. One point is right at the beginning of the planning stage: if the discussion about sustainable energy and resources management is carried out in a transparent, open and participatory process, further steps are much more likely to be accepted by all parties and will help prevent conflicts at later project stages. According to the WCD there are several fundamental strategic points in the decision making process. One point is right at the beginning of the planning stage: if the discussion about sustainable energy and resources management is carried out in a transparent, open and participatory process, further steps are much more likely to be accepted by all parties and will help prevent conflicts at later project stages.

The strategic priority 'Addressing existing dams' includes (among other policy principles) the identification and assessment of outstanding social issues associated with existing large dams. Considering that between 40 and 80 million people have been displaced by large dams (a lot more have been directly or indirectly affected) this is a difficult task. Financial institutions, development agencies and companies are a lot more interested in looking at the future than dealing with the complex problems of the past. But there is no way out. Dam-affected people of the past along with future dam-affected people are unlikely to accept new dams and believe in what is being promised while the legacy of the past is being forgotten. To solve these problems is a condition for further constructive discussions if not a mere moral obligation. How can those responsible for planning future dams think of gaining public acceptance if not by showing that they mean what they say?

The recommendations of the WCD are of fundamental importance, because they were agreed upon in a multi-stakeholder process. In other words, they were made by representative of industry as well as those from dam-affected people, by members coming from industrialized countries as well as those from developing countries. Within the Commission, it was possible to integrate the most diverse views and to come to a consensus.

***No More Dams?***

Although in their assessment the evidence would have allowed even stronger recommendations, NGOs and people's movements welcomed the report and asked for immediate implementation. Other stakeholders do not seem to be very enthusiastic about it. Quite a few representatives from industry and investors seem to fear that the implementation of the WCD recommendations would lead to an abrupt stop in dam building.

All appreciate that the WCD report vindicates many concerns raised by NGO campaigns. Given the role of financial institutions in funding large dams and in the WCD process, and based on the WCD report's recommendations, all has to call on all public financial institutions, including the World Bank, the regional development banks, the export credit agencies an bilateral and agencies, to take the following actions:

- All public financial institutions should immediately and comprehensively adopt the recommendations of the World Commission on Dams, and should integrate them into their relevant policies, in particular those on water and energy development, environmental impact assessment, resettlement, and public participation. In particular, as recommended by the WCD, no project should proceed without the free, prior and informed consent of indigenous people, and without the demonstrable acceptance of all those who would be affected by the project.

- All public financial institutions should immediately establish independent transparent and participatory reviews of all their planned and ongoing dam projects. While such reviews are taking place, project preparation and construction should be halted. Such reviews should establish whether the respective dams comply, as a minimum with the recommendations of the WCD. If they do not, projects should be modified accordingly or stopped altogether.

- All institutions which share in the responsibility for the unresolved negative impacts of dams should immediately initiate a process to establish and fund mechanisms to provide reparations to affected communities that have suffered social, cultural and economic harm as a result of dam projects.

- All public financial institutions should place a moratorium on funding the planning or construction of new dams until they can demonstrate that they have complied with the above measures.

# 8

## Water: *An Educational and Informative Approach*

The most characteristic element of our planet is undoubtedly water. Indeed, more than two-thirds of the earth's surface is covered by water—the total volume representing almost 1,500 million cubic kilometers. About 94 per cent of this water is found in the oceans, almost 6 per cent is located underground and in glaciers whereas rivers, lakes, soil moisture and atmospheric vapour, which constitute the major source of drinking water, account for a mere 0.0221% of the total volume.

Water is indispensable for all living organisms. Life, as we know it, is impossible without water. It is present in all aspects of our life—directly or indirectly next of the air we breathe and together with the soil that we live upon, water constitutes the most important part of our environment, our most precious resource. And yet, except in the arid or semi-arid regions of the world, its value is generally overlooked until some catastrophe—natural or man-induced—forces our attention to its worth. But even so, no sooner is the situation remedied than, more often that not, we revert to our old attitude.

The reason for this sort of indifference is undoubtedly attributable to the fact that, except in exceptional circumstances, water has always been considered as a "gift of the gods", as some thing that human beings are as naturally entitled to as the air they breathe. Its supply, however uneven, has always seemed inexhaustible because water has a natural regenerative cycle which, until the present century, was beyond human control or interference—or even proper comprehension. But the trend of social, political and economic evolution, notably in the past 200 years, with an increase of industry, agriculture, technology, and above all, a vertiginous population growth, as led to a dramatic revision of the age-old belief that no demands made by human populations on the natural resources of the planet are in the process of setting in motion vicious circles in the environment from which it is becoming increasingly difficult to extricate ourselves, not only as concerns the present, but far more important, for the future. Thus, the overuse-or abuse—of water resources has started affecting seriously not only the water cycle but the very nature of water in such a way that, in conjunction with other abuses of the environment, the results have been climate changes, droughts, flooding, desertification on the one hand and acid rain, water pollution and eutrophication on the other.

Actually, the problem of water is to be considered less in terms of quantity—though with a steeply increasing world population making increasingly heavier demands on a fixed quantity of water, one will sooner or later be confronted with this aspect of the problem too—than in terms of proper distribution of available resources taking into account sound management, stock-age and maintenance of quality. For among the major preoccupations of humanity in the coming years, adequate supply of freshwater to the teeming populations figures in the forefront. Between 1900 and 2000 water consumption will have globally increased tenfold and though the share of agriculture, the major consumer of fresh water, is expected to drop significantly (from 90 per cent to 62 per cent approximately), that of industry and the cities will have increased enormously (approximately, from 6 per cent to 24 per cent and 3 per cent to 8 per cent respectively).

Given the current trend of societal evolution i.e. greater emphasis on industry and increasing migration towards the cities added to the global population boom, these figures are certainly cause for concern. Not only because of the damages caused to freshwater resources through the increasing use of fertilizers in the search to maximize agricultural production to cater to the increasing populations, but equally because the mushrooming of industries and urban concentrations are sources of increasing water pollution. Though the industrialised nations have more than their fair share of blame in this matter insofar as the current state of water pollution goes, for the future, it is in the developing world that lies the major source of concern. Lack of resources for adequate urban planning, the increasing role of industry in the search for economic solutions added to uncontrollable population pressures are already on they way to creating an explosive situation in a great number of developing nations with the available water supply becoming more and more inadequate in terms of quantity as well as quality. And then one considers he fact that around 80 per cent of all diseases are estimated to be water related, and that by the year 2000, 51 per cent of the world population will be urban based, one can hardly be accused of exaggeration in speaking of an explosive situation.

Attacking such a vast problem is no mean task. Water being at the very source of life, what concerns water concerns every aspect of life. Thus, be it climate change, pollution, desertification, deforestation, food production....or whatever other major environmental problem that humanity is confronted with today, water constitutes one of the prime factors. Managing our water resources with care and intelligence for the use of present and future generations is a major responsibility which has to be shared by governments and the public alike, for no sector alone can deal efficiently with so vital a problem which affects not only the present but also the future of humanity. Again, as in the case biodiversity and climate change, the problem of water being a global problem, international cooperation is of utmost importance since activities in one part of the planet are likely to produce consequences in other regions of the world. Concerted action by the international community alone is capable of dealing effectively with a problem of such far-reaching consequences.

If our planet is to be saved from disaster—for in jeopardizing our water resources we are guilty of nothing less than condemning life itself on our planet—we have to work for sustainable results: short-term plans for the present which will dovetail into medium-term ones for the coming generations without compromising the possibility, at the same time, of careful long-term planning to guarantee the future of the planet. In this, the part of environmental education and information of the people is fundamental. No strategy, no policy, no plan—be it ever so well prepared and implemented—can hope to succeed without the active and effective participation of the main actors—the people who must be properly educated and informed. For this age-old techniques, beliefs-mentalities must be brought in line with present day realities. People have to learn to think differently in order to veer from a course which, however right in the past, has been shown to be less than adequate for present conditions—and catastrophic for the future—and must therefore needs be altered.

Changing mentalities is neither an easy nor a rapid process. It is difficult to go back upon the accumulated experience of generations—even in the face of stark realities and scientific evidence. Moreover, when dealing with such global and fundamental issues as water, where even "scientific evidence" tends to be stated in tentative terms, the task becomes more onerous. Add to this the fact that the problem presents itself most presently in developing countries which are equally subject to enormous economic pressures which tend to reduce the cope of possible solutions. We are thus faced with the enormous task of trying to change attitudes, values, mentalities of populations whose geographical, socio-cultural and economic conditions have already fashioned priorities other than those that would precisely permit them to overcome their difficulties in a sustainable manner. In other words, of persuading people of abandon traditional short-term strategies in favour of perhaps more unattractive but eventually sustainable, long-term practices.

A veritable Herculean labour—which can only be accomplished through information and education. And in particular, through environmental education and information whose avowed aim is precisely to develop the understanding, knowledge, skills and motivations leading to the acquisition of attitudes, values and mentalities which are necessary to deal effectively with environmental issues and problems. Sound and systematic environmental education of the people associated with concerted local, national and international action, is the only means to finding a sustainable solution to this problem. The ground has to be paved through adequate information on the subject followed by educative processes adapted to specific local conditions. For a uniform education, whether formal or non-formal, might perhaps do more harm than good as its rejection, due to its unsuitability in the light of local customs, beliefs, traditions.., might only serve to reinforce the very attitudes that it seeks to change. In each region, each country, each locality the educative processes must correspond to the socio-cultural, historical, economic conditions of the people. Only then can we hope to arrive at the change in mentalities around the planet which, coupled with consistent, parallel support from national and international institutions, will lead to the safeguard of what is perhaps our most precious resource—Water.

9

# Population Growth and Fresh Water

Wherever population is growing, the supply of fresh water per person is declining. As a result of population growth, the amount of water available per person from the hydrological cycle will fall by 74 per cent between 1950 and 2050. Stated otherwise, there will be only one fourth as much fresh water per person in 2050 as there was in 1950. With water availability per person projected to decline dramatically in many countries already facing shortages, the full social effects of future water scarcity are difficult even to imagine. Indeed spreading water scarcity may be the most underrated resource issue in the world today.

Evidence of water stress can be seen as rivers are drained dry and as water tables fall. The Colorado River in the south-western United States now rarely reaches the sea. The Yellow River, the northernmost of China's two major rivers, has run dry for a part each year since 1985, with the dry period becoming progressively longer. In 1997, it failed to make it to the sea for 226 days. The Nile, the largest river in the Middle East, has little water left when it reaches the sea.

Water tables are now falling on every continent, including in major food-producing regions. Among those where aquifers are being depleted are the U.S. southern Great Plains the North China Plain, which produces nearly 40 per cent of China's grain; and most of the India. Wherever water tables are falling today, there will be water supply cutbacks tomorrow, as aquifers are eventually depleted.

Some 70 per cent of the water pumped from underground or diverted from rivers is used for irrigation, 20 per cent is used for industrial purposes, and 10 per cent is for residential use. Water use patterns vary widely by region. In Europe, for example, where agriculture is largely rainfed, water withdrawals are dominated by industrial use. In Asia, in contrast, irrigation accounts for 85 per cent of all water use.

As countries press against the limits of their water supplies, the competition among sectors itensifies The economics of water use does not favour agriculture. One thousand tons of water can be used to produce one ton of wheat worth $200 of to expand industrial output by $14,000. This ratio of 70 to 1 explains why industry almost always wins in the competition with agriculture for water.

As the growing demand of water collides with the limits of supply, countries typically satisfy rising urban and residential demands by diverting water from irrigation. They then import grain to offset the loss of irrigation water. Since it takes at least 1,000 tons of water to produce a ton of grain, importing grain becomes the most efficient way to import water. North Africa and the Middle East—a region where population growth is rapid and every country faces water shortages—has become the world's fastest growing grain import market during the 1990s. In 1997, the water required to produce the grain and other foodstuffs imported into the region was roughly equal to the annual flow of the Nile River.

In both China and India, the two countries that together dominate world irrigated agriculture, substantial cutbacks in irrigation water supplies lie ahead. The combination of the effects of aquifer depletion in key countries such as these and the growing diversion of irrigation water to nonfarm uses in many countries makes it unlikely that there will be much, if any, increase in total irrigated area over the long term. Already the irrigated area per person has been slowly declining since 1978, falling from a historical high of 0.047 hectares per person to 0.045 hectares in 1996—a drop of 4 per cent. If the total irrigated area remains at roughly 263 million hectares until 2050, this key figure will fall to 0.028 hectares per person in 2050—declining by an additional 38 per cent. Such a shrinkage will pose a formidable challenge to the world's farmers.

About a billion people will be living in countries facing absolute water scarcity by 2025. These nations do not have enough water to maintain 1990 levels of food production per person from irrigated area, even with high irrigation efficiency, and to meet the needs for domestic, industrial, and environmental purposes as well. They will have to reduce water use in agriculture in order to satisfy residential and industrial water needs. The resulting decline in domestic food production will force them to import more food, assuming it is available. Although detailed water projections by sector for each country are not available for 2050, the number of water-deprived people will be far greater than in 2025 if the world continues on the U.N. medium population trajectory. The bottom line is that if we are facing a future of water scarcity, then we are also facing future of food scarcity.

10

## South Asia Quarrels Over Water

Bangladesh, Bhutan, India and Nepal have at least three features in common; they are all situated on the southern slopes of the Himalayas, they have high levels of poverty and they are all rich in water resources. To these add a fourth: they frequently quarrel about water.

For decades, policy makers in the four countries have grappled with the problem of how to harness the rivers that flow from the mountains and serve the region's over-one billion people. Mostly, they have gone in for large dams and grand irrigation schemes. But though these promise a lot, their usefulness is being increasingly questioned by many independent engineers, scientists and social activists. And they are thought to contribute to water conflicts between countries.

Large hydro-power projects represent a development model that is biased toward industry and urban areas. And irrigation schemes favour rich farmers with large farmlands, most useful for water-and fertilizer-intensive cash crops that poor peasants cannot afford to grow. Because of their size, they also cause huge displacements-again, of poor farmers. Social activists say that while such projects may have increased overall food production, they have also increased the rich-poor gap in a region that is home to the world's largest number of absolute poor-people living on incomes of less than one dollar a day.

In tandem, many South Asian hydro-power engineers and economist say 'national' and top-down engineering solutions to water management ought to be replaced by a new 'basin-wide' or regional approach through environmental and socially benign projects.

One example of such 'national solutions' is the Indian Farrakka Barrage on the Ganges. Built near the Bangladesh border in order to divert water to the Calcutta port, it has led to problems with falling water tables and salinity downstream in Bangladesh. And although India and Bangladesh resolved the long-simmering dispute in 1996, Dhaka is now proposing another expensive Ganges Barrage to solve the problems created by the first barrage.

Of the four countries, Nepal has the highest per capita potential for hydropower generation—its narrow valleys and steep terrain mean water flows faster, which is good for power generation. But although Nepal could potentially turn its rivers into 'hydrodollars', Himalayan dams are expensive to build and the country's experience in dealing with India on joint river projects has been patchy. At the same time exporting power to India, where powercuts that can last up to 12 hours at a time in some cities, means foreign exchange for Nepal.

Because many Nepalese perceive past border irrigation projects as unfairly benefiting India, joint Indo-Nepal river projects are a political issue. A recent treaty to harness the Mahakali River on Nepal's western border with India is a case in point.

Although India and Nepal agree on equally sharing the 7,500 megawatts of power from the project, negotiations have been stalled over Nepal's demand that it be compensated for downstream irrigation benefits to India. Nepal is not allowed to harness water in many of its rivers for its own use without India's concurrence these rivers flow into India, and Nepal is bound by past bilateral treaties.

As one official at Nepal's Water and Energy Commission put it: "What's in it for Nepal? Why should we write off benefits from the Mahakali for future generations of Nepalis? Thing look even bleaker for larger joint projects like a planned 10,800 mw dam, on the Karnali River the $ 10 billion Chisapani dam project is expected to displace some 60,000 Nepali farmers. In addition, India has already used up all the natural flow of the Karnali for its own irrigation downstream.

Things are different when it comes to Bhutan: India and the tiny Himalayan kingdom sorted out their river projects without any major difficulty although experts say it is too early to assess how much the Bhutanese will really benefit. According to an estimate by an Indian water consultant, Bhutan can generate up to 20,000 mw of power from its rivers.

Generating funds for large projects does not seem to be a problem. Traditionally, the World Bank has been one of the major financiers for dams worldwide having undertaken 400 projects involving dams since 1970. In 1988, India alone had about 473 dams under construction, with the Bank involved in 45 of them.

When the World Bank recently pulled out of some controversial mega projects, such as the Arun 111 dam in Nepal, multinational companies stepped in. In Nepal, three medium-sized hydropower projects worth more than 300 million dollars are being built with foreign direct investment. FDIs are backed by organisations like the Japanese-managed Global Infrastructure Fund (GIF) which is interested in putting money into such projects as the 270 metre-high Kosi High Dam in Nepal, the Ganges Barrage in Bangladesh and the mammoth 20,000 mw Dhihang dam in India. Funds also come from the Manila-based Asian Development Bank.

But many resource economists maintain that the chances of making colossal mistakes is higher with large projects. It is all a question of risk-management. Can our countries afford to take the risk of spending billion on doubtful projects that can go dangerously wrong?

11

## Water: *Will Be There Enough?*

Humankind has a special relationship with water. In every civilisation, the most ancient traditions associate this precious resource with the origins of life, purification and regeneration. Far from being a mere raw material such as oil, water is vital for life, indispensable to the economy and so rich in symbolic value that it triggers passionate responses. All the computers in the word will never be able to express the real perception of the value of water or codify the interactions between it and peoples.

For decades, experts have been making grim forecast that the Earth will start running out of water and that conflicts over this precious liquid will erupt into wars. The situation is indeed alarming.

How much water is there in the earth's reserves? Highly expensive probes have been sent to the moon, mars and the satellites of Jupiter and Saturn to findout whether there is water on them, but we still lack accurate data about the earth's hydrological resources. Such information would help to provide a cleaner picture of the future and, especially to foresee the global repercussions of demographic and climate change.

One thing we know about water is that there is plenty of it. The total volume is put 1.4 billion cubic Kilometers—which could be imagined as a 2,650-metre-deep layer of liquid evenly disturbed over the entire surface of the planet. But 98 per cent of it is salt water, mainly in the oceans and seas. Most of the earth's fresh water is trapped in the polar ice caps, less than 1 per cent of it is available in lakes, rivers and shallow, easily-accessible acquifers. These water resources are constantly in flux. Water from the oceans and land evaporates into the atmosphere before falling again as rain or snow, nourishing plants and swelling rivers that flow into the sea. *It also seeps* through the ground and percolates down to acquifers. Very deep groundwater, known as fossil water is impervious to seepage and not renewable.

In the industrialised countries, all you have to do is turn a tap and before you know where you are you've used a considerable amount of water up to 600 litres per person a day in the United States. In hot developing countries, where shanty-towns on the edges of cities are crowded with growing numbers of migrants from the rural areas, a spigot and two litres of water a day are a luxury.

Over 1.3 billion people received improved drinking water services and some 750 million got better sanitation facilities during the International Drinking Water Supply and Sanitation Decade (1980-1990). Approximately 1.2 billion people still have no access to drinking water and 2.9 billion lack sanitation. The resulting water-borne diseases take the lives of five million people a year, most of them children.

Farming and manufacturing account for most of the world's water consumption, far outdistancing human needs of the 3,240Km$^3$ of fresh water drawn every year, only 8 per cent are used for human consumption. Each year fewer than ten countries use 60 per cent of the world's 40,000 billion m$^3$ of surface and ground water. Lastly, per capita consumption rises with the standard of living, ranging from 260 litres a day person in Israel to 200 in Europe, 70 for a Palestinian on the West Bank and 30 in Africa.

**The Dangers of Irrigation**

Demand runs highest in places where irrigation is indispensable, such as central Asia, Iraq, Iran, Pakistan

Madagascar and also in some industrially developed countries such as the United States. Farming accounts for two-thirds of the total water resources used by humans—a figure that rises to 80 per cent in the Southern countries. Developing countries consume twice as much water per hectare to irrigated land as industrialised nations, yet their production is three times lower.

Because of the heat, half the water, evaporates in storage areas or when flowing through open-air irrigation canals, poorly conceived irrigation rejects lead to deterioration of the soil.

The first is Pakistan, during the first half of the twentieth century, 10 million hectares were abundantly irrigated in the Indus plain. Waterlogging caused by irrigation, combined with a high rate of evaporation, has led to salinisation of the soil, making it unproductive. The second example, the Aral Sea in the former Soviet Union, is different but the result is the same. Much of the water from the Syrdarya and Amu Darya rivers that flow into the huge lake has been diverted to feed 1,80,000 Kms of irrigation canals, only 12 per cent of which have been made watertight. The rivers' flow is considerably restricted and the Aral Sea is drying up. Irrigation for agricultural purposes is expensive. To make it profitable farmers must use massive amounts of pesticides, herbicides and fertilizer on increasingly exhausted soil. The impact of pesticides on health has been overlooked. The child morbidity and mortality rates are among the world's highest.

12

## Water Problem in South India

Southern India's fast-growing urban areas and its farmers will collide over water allocation unless the government and water users take swift action. Urgently needed are measures that ensure efficient water use in cities and on farms, including regulation of groundwater withdrawal, restoration of traditional rain-collecting reservoirs, and experimentation with different cropping strategies.

In India's arid southern tip, typifies the plight of the country's extensive drylands. Scanty rainfall means residents often must get by on water drawn from small reservoirs and underground sources.

Extended dry periods are punctuated by intense monsoon bursts. When rains lash South India during the monsoons, only a fraction of the downpour is captured for later use. The rivers are seasonal and small compared to the Himalayan cataracts up north.

During the monsoon, these waters swell briefly to gigantic proportions recharging the water tables in their basins and then subside. In times of drought, when the rivers are narrow ribbons, water drawn from below ground sustains crops.

In the cities of South India the problems of water supply are exacerbated by antiquated or nonexistent infrastructures that cannot keep up frentic growth.

Farmers in South India depend on irrigation to see them through the growing season. They draw their water from three sources: Government canals that bring water from rivers and dams, traditional reservoirs known as "Tanks", and public and private wells often fitted with electric pumps. All three need to be made more efficient.

Besides being enormously expensive, large surface irrigation canals and dams are plagued by massive leaking and evaporation. Tanks collect rain water runoff behind small earthen dams. But these are falling into disrepair as farmers take advantage of low interest government loans and heavily subsidised electricity to switch to private wells fitted with electric pumps. Without maintenance, the tanks fill up with silt and hold less water.

As more wells are dug and as percolation tanks that once recharged underground water supplies fall into disuse, the water table is dropping at a rapid rate. Farmers who can afford to install powerful electric or diesel pumps on their wells are able to tap the retreating water table, but poorer farmers who raise their water by hand or cattle power from shallow wells often come up dry.

No regulations govern the amount of water a farmer can withdraw from a well, so underground reserves are available on a first-come, first-served basis. Add to this situation electricity rates based on the horsepower of pumps, not on the amount of electricity they consume, and there' no incentive to save water.

Any effort to limit groundwater extraction carries heavy political liabilities. Politicians are reluctant to risk raising the ire of farmers, who provide the majority of their votes.

Preventing overdraft where water tables are falling would be a first step toward correcting water troubles. While some areas face the prospect of rationing, others have an abundance of groundwater still to be tapped. Effective legislation must be based on detailed and accurate map of groundwater supplies—something that doesn't currently exist.

They should also consider instituting higher user fees for its public canal projects and retuning to its old policy of charging farmers according to the electricity they consume. Making farmers pay closer to the full price for water and electricity would provide revenue for digging communal wells equipped with electric pumps and overhauling the tank system, while providing an incentive for farmers to conserve.

Using tanks and wells together maximizes the effectiveness of both, since tanks take pressure of groundwater supplies and wells sustain crops during their final weeks of growth when tanks are low. Any efforts to limit the number of new wells dug has to include measures that provide something to those without water, otherwise only those who currently own wells will benefit from groundwater conservation.

Officials might also experiment with encouraging farmers to plant less water-demanding crops such as ragi, sorghum and pulses in place of cotton, sugarcane, bananas and spices.

If politicians act now, they still have a chance to craft measures that are fair rather than desperate. The time is nearing water for drinking and irrigating crops will have to be called from careful conservation rather than from the ground.

13

# Population Growth and Grain Production

The relationship between the growth in world population and the grain harvest has shifted over the last half-century, neatly dividing this period into two distinct eras. From 1950 to 1984, growth and the grain harvest easily exceeded that of population, raising the harvest per person from 247 kilograms to 342, a gain of 38 per cent. During the 14 years since then, growth in the grain harvest has fallen behind that of population, dropping output per person from its historic high in 1984 to an estimated 317 kilograms in 1998—a decline of 7 per cent, or 0.5 per cent a year.

These global trends conceal widely divergent developments among countries, contrasts that can be seen for the world's two most populous nations: India and China. In both, grain production per person was close to 200 kilograms as recently as 1978. Since then, the figure in India has edged up slightly but still falls short of 200 kilograms, while in China production has surged since the economic reforms in 1978, with per-person output now at nearly 300 kilograms. The combination of a dramatic surge in grain production and an equally dramatic reduction in population growth has given China a large margin of safety, effectively eliminating most of its hunger and malnutrition. Meanwhile, although India has also achieved impressive gains in its harvest, these have been largely cancelled by population growth, leaving its 976 million people living close to the margin.

What has happened in China and India is the story of developing countries in general. The overwhelming majority have achieved substantial, if not dramàtic, gains in their grain harvests over the last half-century. Some, such as Thailand, have combined this with a much slower growth of population, which means that agricultural gains translate into rising grain production per person. In Pakistan, by contrast, grain production per person climbed steadily for a while, but it peaked in 1981 at 186 kilograms. Since then it has been declining nearly 1 per cent a year. In effect, Pakistan's farmers are losing the battle with population growth.

The slower growth in the world grain harvest since 1984 is due to the lack of new land and to slower growth in irrigation and fertilizer use. Irrigated area per person, after expanding by 4 per cent since then as growth in the irrigated area has fallen behind that of population.

The increase in world fertiliser use has slowed dramatically since 1990, as diminishing returns to the application of additional fertilizer has stabilized use in the United States, Western Europe, and Japan and slowed annual growth in world fertilizer use from 6 per cent between 1950 and 1990 to scarcely 2 per cent in recent years.

Although Malthus was primarily concerned with the additional demand for grain generated by population growth, rising affluence is also playing a role. In a low income country such as India, grain consumption per person is less than 200 kilograms per year and diets are typically dominated by a single starchy staple-rice, for instance. With scarcely a pound of grain available a day per person, nearly all must be consumed directly, leaving little for conversion into animal protein. For the average American, on the other hand, the great bulk of the 800-kilogram daily grain consumption is taken in indirectly in the form of beef, pork, poultry, eggs, milk, cheese, ice cream, and yogurt. At the intermediate level, in a country like Italy, people consume 400 kilograms of grain a day. Future food price stability thus depends on expanding production fast enough to keep up with both population growth and rising affluence.

One question often asked is, How many people can the Earth support? This must be answered with another question, At what level of consumption? If the world grain harvest of 1.87 billion tons were expanded to 2 billion tons in the years ahead, it would support 10 billion Indians or 2.5 billion Americans. To answer the question of how many people the Earth can support, we first have to know the level of consumption we expect to live at.

Now that the frontiers of agricultural settlement have disappeared, future growth in grain production must come almost entirely from raising land productivity. Unfortunately, this is becoming more difficult. After rising at 2.1 per cent a year from 1950 to 1990, the annual increase in rainland productivity dropped to scarcely 1 per cent from 1990 to 1997. The challenge for the world's farmers is to reverse this decline at a time when cropland area per person is shrinking, the amount of irrigation water per person is dropping, and the crop yield response to additional fertilizer use is falling.

14

## Fresh Water and the Environment

It is widely recognized that water is going to be one of the major issues confronting humanity at the turn of the century and beyond. We are facing a crisis as regards the quantity and quality of water supply, but we have yet to experience full social and political impact of that crisis. The escalation in the population and the quest for continued development is leading to conflicting pressures on water resources. Such resources are the ultimate recipient of pollution from various socio-economic activities associated with urbanisation, agriculture, mining and clearing of native vegetation. Pollution originating from human waste, especially where appropriate sanitation facilities are not available, or are located too close to water supply sources affects both surface water and ground water.

This makes water supply and health perhaps the most important issue for the large proportion of the global population. Paradoxically, the demands for "sustainable management" and increasing global population require more potable water from a declining available potable water base.

It is universally accepted that proper water administration is a critical component of sustainable development—that is, development that meets the needs of both present and future generations. Indeed, water is an essential factor in a large number of productive activities, of which one of the most important is the production of food by irrigation. This activity, accounts for two thirds of the water resources used by humanity. A supply of drinking water and sanitation in urban centres are crucial for preserving human health.

For some decades it has been known that the misuse of water resources is responsible for many important environment problems. For example, in many industrialized cities both surface water and ground water are seriously contaminated. This deterioration is a consequence of a range of human activities, sometimes in isolation, others over a large area or a long period of time. Among examples of the latter is modern agriculture, whether it uses irrigation or not, as a result of the intensive use made of mineral fertilizers and pesticides.

**Water Shortage: Exaggeration, Reality or Bad Management?**

Some of these problems have made news and have created the impression the water shortage will be one of humanity's big problems in the coming decades. Sometimes this feeling is due to genuinely manipulative publicity campaigns to justify the setting in motion of hydraulic megaprojects which basically benefit a few large construction companies. The truth is that except for a handful of very specific cases, no problems of water shortage are to be found almost anywhere. On the other hand, cases of bad water management are not rare at all.

**Basic Principles for Good Water Management**

Good management of water resources—and of almost all other natural resources—must be based on the principles of solidarity, "subsidiarity" and participation. The physical reality requires that these resources be considered a common heritage of humanity both now and in the future. By "subsidiarity" we mean that water management should be as decentralized as possible: what one person or any minor social group can do should not be done by a higher authority. For example, what local government can do should not be done a regional, state or central government. Participation consists in water users playing as large a part as possible in decisions affecting water, in keeping with each state's or country's social and cultural structure. Obviously this participation calls for a certain cultural and technical knowledge—a hydrological education—on the part of those users.

The need for participation by users is even greater in the exploitation of groundwater. In this case, users tend to extract water independently of one another. They often fail to realize, until there is a serious economic or environmental impact, that their pumping affects other people who rely on the same water supply as has happened.

Water shortage is rarely a serious problem: in fact, in some cases the problem is exaggerated to justify the construction of large works using taxpayer's money. On the other hand, the contamination of surface and groundwater tends to be a problem which rarely receives adequate treatment. Successful water management should be based on three basic principles: solidarity, subsidiarity and participation. The specific way in which these principles are applied will vary from one state or country to another, but the effectiveness of water management will depend in large measure on the hydrological education of the general public.

The universal way of obtaining freshwater is from rain. River systems are the results of the excess water that falls on dry land in the form of rain. On the one hand, rainwater penetrates the permeable soils, saturates them and accumulates to form groundwater reservoirs, or aquifers, which can come to the surface in the form of springs. On the other hand, the water is absorbed by vegetation, which uses it for pumping minerals and then evaporates it by transpiration. Some rainwater is lost because it evaporates immediately on falling on impermeable surfaces like the asphalt of roads and cities. Running water courses finally flow over saturated soils, shaping the complex systems of the watersheds or river basins.

Since each basin's natural system has developed gradually and has grown up according to the yearly distribution and fluctuations of rainfall, we have to appreciate that any large-scale project for redistributing water by means of pipes, as if it were gas or electricity, is a journey into the unknown. This is because it destroys the results of the work of shaping the climate, however transitory it might be.

**Variable Volumes**

All water supplies are of variable volume. Both the discharge of rivers and the level of lakes and aquifers depend on rainfall. As these resources are components of a larger system, the river basin, a reasonable policy would be to manage water resources according to the characteristics of each basin. This would require, first of all, a proper understanding of the system so as to adapt use and consumption to the existing supply. Conserving river systems as much as possible in their natural sate is the best guarantee for the preservation of the landscape and of a constant supply. Grondwater reservoirs aren't canals, but are more like lakes, so that pollution leads to the build-up of a debt which is paid in years to come.

**Consumption**

Water consumption has increased in recent years as a result of not only population growth but also an increase in living standards. In the rural areas the introduction of new farming methods, the spread of irrigation and the excessive use of fertilizers and pesticides causes very high consumption—it is estimated that more than 2/3 of water consumption is used in irrigation. Agricultural pollution also endangers both surface water and aquifers, which receive water full of chemical products. Many cases of eutrophication, the enrichment of water by nutrients that accelerate the growth of algae, derive from the run-off of fertilizers. The practice of intensive stock-raising on farms with large numbers of animals also brings about these problems of over consumption and pollution. Cleaning the stockyards requires large amounts of water which is then released into the environment with high concentrations of nitrogen.

As for industries, they have in the past taken little care over water consumption and dumping, and in many areas the need for proper attention comes as something new. The best thing would be to make industry take its water at a point down-river from where it returns it or, better still, generalise the use of closed circuit systems based on the constant recycling and reusing of the same water.

As regards human consumption, the general attitude to cleanliness is based on diluting pollutants. One example is the success of the use of the Water Closet which involves diluting a few decilitre of urine in 10 or more litres of drinking water: quite a record in wastefulness.

Another aspect to be considered is the different quality of the water that falls on well formed soils from the water that falls on roads, cities, airports, suburbs and built-up areas and whose composition is less stable and "worse" than that resulting from a more uniform interaction with mature soils. Remember that streets, roofs, communication routes, airports and built-up areas already cover a high proportion of the earth's land area and are still on the increase.

Purification techniques should be based especially on the natural processes that include biological activity. Otherwise—for example, if physico-chemical methods are used—there can be side-effects such as an excess of mud or sediments. The strategy to follow is to optimize operations in our use of water according to the discharge and to the distribution of contamination. A system in the form of a conduit or channel, such as a river, can respond relatively quickly. On the other hand, lakes and dams can only do so up to a point, because they show more inertia and irreversibility and take longer to clean.

Large lakes, not to mention the sea, might seem a good place to dump contaminating refuse, but they can't then be cleaned. This is the price we pass on the future generations: a comfortable attitude, but an unacceptable one.

15

## A Rare and Precious Resource

Fresh water is a scarce commodity. Since it's impossible to increase supply, demand and waste must be reduced. But how?

Water is a bond between human beings and nature. It is ever-present in our daily lives and in our imaginations. Since the beginning of time, it has shaped extraordinary social institutions, and access to it has provoked many conflicts.

But most of the world's people, who have never gone short of water, take its availability for granted. Industrialists, farmers and ordinary consumers blithely go on wasting it. These days, though, supplies are diminishing while demand is soaring. Everyone knows that the time has come for attitudes to change.

Few people are aware of the true extent of fresh water scarcity. Many are fooled by the huge expanses of blue that feature on maps of the world. They do not know that 97.5 per cent of the planet's water is salty—and that most of the world's fresh water—the remaining 2.5 per cent—is unusable: 70 per cent of it is frozen in the icecaps of Antarctica and Greenland and almost all the rest exists in the form of soil humidity or in water tables which are too deep to be tapped. In all, barely one per cent of fresh water—0.007 per cent of all the water in the world, is easily accessible.

Over the past century, population growth and human activity have caused this precious resource to dwindle.

Between 1900 and 1995, world demand for water increased more than six fold-compared with a threefold increase in world population. The ratio between the stock of fresh water and world population seems to show that in overall terms there is enough water to go round. But in the most vulnerable regions, an estimated 460 million people (8 per cent of the world's population) are short of water, and another quarter of the planet's inhabitants are heading for the same fate. Experts say that if nothing is done, two-thirds of humanity will suffer from a moderate to severe lack of water by the year 2025.

Inequalities in the availability of water—sometimes even within a single country—are reflected in huge difference in consumption levels.

Scarcity is just one part of the problem. Water quality is also declining alarmingly. In some areas, contamination levels are so high that water can no longer be used even for industrial purposes. There are many reasons for this—untreated sewage, chemical waste, fuel leakages, dumped garbage, contamination of soil by chemicals used by farmers. The worldwide extent of such pollution is hard to assess because data are lacking for several countries. But some figures give an idea of the problem. It is thought for example that 90 per cent of waste water in developing countries is released without any kind of treatment.

Things are especially bad in cities, where water demand is exploding. For the first time in human history, there will soon be more people living in cities than in the countryside and so water consumption will continue to increase. Soaring urbanisation will sharpen the rivalary between the different kinds of water users.

**Curbing the Explosion in Demand**

Today, farming uses 69 per cent of the water consumed in the world, industry 23 per cent and households 8 per cent. In developing countries, agriculture uses as much as 80 per cent. The needs of city-dwellers, industry and tourists are expected to increase rapidly, at least as much as the need to produce more farm products to feed the planet. The problem of increasing water supply has long been seen as a technical one, calling for technical solutions such as building more dams and desalination plants. Wild ideas towing chunks of icebergs from the poles have even been mooted.

But today, technical solutions are reaching their limits, Economic and socio-ecological arguments are levelled against building new dams, for example: dams are costing more and more because the best sites have already been used, and they take millions of people out of their environment and upset ecosystems. As a result, twice as many dams were built on average between 1951 and 1977 than during the past decade.

Hydrologists and engineers have less and less room for manoeuvre, but a new consensus with new actors is taking shape. Since supply can no longer be expanded—or only at prohibitive cost for many countries—the explosion in demand must be curbed along with wasteful practices. An estimated 60 per cent of the water used in irrigation is lost through inefficient systems, for example.

Economists have plunged into the debate on water and made quite a few waves. To obtain "rational use" of water i.e. avoiding waste and maintaining quality, they say consumers must be made to pay for it. Out of the question, reply those in favour of free water, which some cultures regard as "a gift from heaven". And what about the poor, ask the champions of human rights and the right to water? Other important and prickly questions being asked by decision-makers are how to calculate the "real price" of water and who should organise its sale.

**The State as Mediator**

The principle of free water is being challenged. For many people, water has become a commodity to be bought and sold.

But management of this shared resource cannot be left exclusively to market forces. Many elements of civil society—NGO's, researchers, community, groups—are campaigning for the cultural and social aspects of water management to be taken into account.

Even the World Bank, the main advocate of water privatisation, is cautious on this point. It recognises the value of the partnerships between the public and private sectors which have sprung up in recent years. Only the state seems to be in a position to ensure that practices are fair and to mediate between the parties involved—consumer groups, private firms and public bodies. At any rate, water regulation and management systems need to be based on other than purely financial criteria. If they aren't, hundreds of millions of people will have no access to it.

16

# Solutions for A Water-Short World

As populations grow and water use per person rises, demand for fresh water is soaring. Yet the supply of fresh water is finite and threatened by pollution. To avoid a crisis, many countries must conserve water, pollute less, manage supply and demand, and slow population growth.

Caught between growing demand for fresh water on one hand and limited and increasingly polluted water supplies on the other, many developing countries face difficult choices. Populations continue to grow rapidly. Yet there is no more on earth now than there was 2,000 years ago, when the population was less than 3 per cent of its current size. Rising demands for water for irrigated agriculture, domestic (municipal) consumption, and industry are forcing stiff competition over the allocation of scarce water resources among both areas and types of use.

Today 31 countries, accounting for under 8 per cent of the world population, face chronic fresh water shortages. By the year 2025, however, 48 countries are expected to face shortages, affecting more than 2.8 billion people – 35 per cent of the world's projected population. Among countries likely to run short of water in the next 25 years are Ethiopia, India, Kenya, Nigeria, and Peru. Parts of other large countries, such as China, already face chronic water problems.

In much of the world polluted water, improper waste disposal, and poor water management cause serious public health problems. Such water-related diseases as malaria, cholera, typhoid, and schistosomiasis harm or kill millions of people every year. Overuse and pollution of water supplies also are taking a heavy toll on the natural environment and pose increasing risks for many species of life.

**What Can Be Done?**

It may already be too late for some water-short countries with rapid population growth to avoid a crisis. Many other countries can avoid the coming crisis if appropriate policies and strategies are formulated and acted on soon. Whether water is used for agriculture, industry, or municipalities, there is much room for conservation and better management. Effective strategies must consider not only managing the water supply better but also managing demand better.

To avoid catastrophe over the term, it also is important to act now to slow the growth in demand for fresh water by slowing population growth. Currently, in many developing countries millions of people want to plan their families and to use contraception. Family planning programmes have played an important role in assuring individual reproductive health and in reducing national fertility levels. Continuing and expanding these programmes also can help assure that population growth eventually slows to sustainable levels in relation to the supply of fresh water.

**Towards a Blue Revolution**

The world needs a Blue Revolution to conserve and manage fresh water supplies in the face of growing demand from population growth, irrigated agriculture, industries, and cities—just as the Green Revolution transformed agriculture in the 1960s. A Blur Revolution will require coordinated responses to problems at local, national, and international levels.

Locally led initiatives show that water can be used much more efficiently. When communities manage fresh water resources efficiently, they also manage other natural resources better, improve sanitation, and reduce disease. At the national level, especially in water-short regions with dense populations, adopting a watershed or river-basin management perspective is a needed alternative to uncoordinated water-management policies by separate jurisdictions. At the international level countries that share river basins can fashion workable policies to manage water resources more equitably. Development agencies need to focus more on assuring the supply and management of fresh water resources and on providing sanitation as part of development and public health programmes.

A water-short world is an inherently unstable world. As the next century dawns, water crises in more and more countries will present obstacles to better living standards and better health and even bring risks or outright conflict over access to scarce fresh water supplies. Finding solutions should become a high priority now.

17

## The Coming Water Crisis

Fresh water is emerging as one of the most critical natural resource issues facing humanity. The world's population is expanding rapidly. Yet there is no more fresh water on earth now than there was 2,000 years ago, when the population was less than 3 per cent of its current size.

Water is, literally, the source of life on earth. The human body is 70 per cent water. People begin to feel thirst after a loss of only 1 per cent of bodily fluids and risk death if fluid loss nears 10 per cent. Human beings can survive for only a few days without fresh water. Yet in a growing number of places people are withdrawing water from rivers, lakes, and underground sources faster than they can be recharged—"unsustainably mining what was once a renewable resource," as one researcher puts it. Currently, 31 countries—mostly in Africa and the Near East—face water stress or water scarcity.

Population growth alone will push an estimated 17 more countries, with a projected population of 2.1 billion, into these water-short categories within the next 30 years. By the year 2025, 48 countries, with more than 2.8 billion people—35 per cent of the projected global population in 2025—will be affected by water stress or scarcity. Another nine countries, including China and Pakistan, will be approaching water stress.

Beyond the impact of population growth itself, the demand for fresh water has been rising in response to industrial development, increased reliance on irrigated agriculture, massive urbanisation, and rising living standards. In this century, while world population has tripled, water withdrawals have increased by over six times. Since 1940 annual global water withdrawals have increased by an average of 2.5 per cent to 3 per cent a year compared with annual population grcwth of 1.5 per cent to 2 per cent. In developing countries over the past decade water withdrawals have been increasing by 4 per cent to 8 per cent a year.

Moreover, the supply of freshwater available to humanity is shrinking, in effect, because many fresh water resources have become increasingly polluted. In some countries lakes and rivers have become receptacles for a vile assortment of wastes, including untreated or partially treated municipal sewage, toxic industrial effluents, and harmful chemicals leached into surface and ground waters from agricultural activities.

Caught between finite and increasingly polluted water supplies on one hand and rapidly rising demand from population growth and development on the other, many developing countries face uneasy choices. The lack of fresh water is likely to be one of the major factors limiting economic development in the decades to come.

**Slowing Demand, Conserving Supplies**

To avoid a water crisis, particularly in water-short counties with rapid population growth, it is vital to slow the growth in demand for water by managing the resource better, while at the same time slowing population growth as soon as possible. Family planning programmes play an important role not only for individual reproductive health but also for sustainability of the use of fresh water and other natural resources in relation to population size.

As population grows, so does demand for fresh water of food production, household (municipal) consumption, and industrial uses. The availability of fresh water limits the number of people that an area can support and affects standards of living. In turn, population growth and density typically affect the availability and quality of water resources in an area, as people attempt to assure their water supply by digging wells, constructing reservoirs and dams, and diverting the flow of rivers. If needs consistently outpace available supplies, at some point overuse of water leads to the depletion of surface and groundwater resources, triggering chronic water shortages.

Scarce and unclean water supplies are critical public health problems in much of the world. Polluted water, water shortages, and unsanitary living conditions kill over 12 million people a year.

Competition for fresh water supplies breeds social and political tensions. River basins and other water bodies do not respect national borders. For example, one country's use of upstream water often subtracts from the supply available for use by downstream countries. As the 21$^{st}$ century dawns, there is a growing risk that wars will be fought over access to fresh water supplies.

If a crisis is to be averted, the world's overuse and misuse of fresh water must end as soon as possible. We cannot afford to keep wasting and fouling our precious supplies of fresh water. Increasingly, human activities are altering the flow of water and drawing down fresh water supplied faster than they can be replenished. Throughout the world enormous amounts of water are wasted due to inappropriate agricultural subsidies, inefficient irrigation systems, leaky municipal pipes, improper pricing of municipal water, poor watershed management, and other imprudent practices. It is time for widespread conservation measures, effective water management policies, and growing attention to assuring fresh water supplies and decent sanitation as part of development and public health projects.

18

## Watershed Development Programme

The Commissionerate of Rural Development is implementing area development programme i.e., watershed development programmes in dry and degraded lands through district agencies—DDP/DPAP/DWMA.

The programme of dry land development in Andhra Pradesh has undergone a major change from 1995-96, with the introduction of new participatory watershed guidelines based on the recommendations of Dr. Hanumantha Rao committee. The main emphasis of the revised guidelines is on active mobilisation and participation of the stakesholders in the programme at all stages—planning, implementation and management. So far, out of 78.20 lakh ha. dry land to be treated by the Rural Development Department, 36.85 lakh ha. is treated/under treatment through 7903 watersheds/projects since 1995-96. The following watershed development programme are being implemented in Andhra Pradesh:

1. Drought Prone Area Programme (DPAP)
2. Desert Development Programme (DDP)
3. Integrated Wastelands Development Programme (IWDP)
4. Employment Assurance Scheme (EAS)
5. Andhra Pradesh Hazard Mitigation (APHM)
6. Rural Infrastructure Development Fund (RIDF).

In order to combat the frequent recurrence of drought and for development of wasteland in the state, watershed programme is being implemented. A massive action plan for the development of all the degraded and wastelands in ten years was launched during 1996-97. Ten-year action plan has been prepared to develop 100 lakh hectares of degraded and wastelands on watershed basis with an outlay of about Rs. 4000.00 cores from 1997-2007 at the rate of 10 lakh hectares every year.

Andhra Pradesh is one of the few states in the country to have an exclusive administrative agency—DPAP/DDP/ DWMA for the implementation of watershed development programme.

**DPAP/DDP**

Since 1995-96, 2966 DPAP and 552 DDP watershed projects are under implementation in Andhra Pradesh covering an area of 14.83 lakh ha. and 2.76 lakh ha. respectively (total = 17.59 lakh/ha).

291 DPAP and 110 DDP watershed projects are under implementation during 2002-03.

During 2003-04, action will be taken to cover more dry areas depending upon the projects to be sanctioned by the Govt. of India.

**IWDP**

Since 1995-96, 842 watersheds have been sanctioned upto December 2002 to develop 4.78 lakh ha. of wastelands in Non-DPAP blocks under IWDP scheme.

108 watersheds have been sanctioned during 2002-03 upto December 2002 as against target of 144 watershed under IWDP scheme.

During 2003-04, action will be taken to cover more dry areas depending upon the projects to be sanctioned by the Govt. of India.

**EAS (Watersheds)**

Since 1995-96, 1884 watershed have been taken up covering an area of 9.42 lakh ha. and component-wise action is being taken to complete them by March 2003.

**RIDF VI**

1244 RIDF watershed projects have been sanctioned to treat an area of 2.63 lakh ha. with Rs. 139.19 crores in 22 districts and we would complete all works by March 2004. The RIDF—VI projects have been sanctioned with 90 per cent as NABARD loan and 10 per cent GOAP share.

**RIDF-VIII (RD and Ohters Sector)**

NABARD has sanctioned 290 SMC projects under RIDF-VIII (RD and Others Sector) to cover an area of 1.48 lakh ha. with Rs. 41.08 crores to 17 districts with 75 per cent NABARD loan of Rs. 30.81 crores and 25 per cent per cent of project cost with rice component of Rs. 10.27 crores during 2002-03.

**APHM and ECRP**

It is a World Bank assisted programme which commenced in July 1997 with a target of 100 watersheds @ 20 watersheds per district. The total area treated is 0.5 lakh ha. The project has been completed in July 2002.

**Neeru-Meeru Programme**

The Government of Andhra Pradesh has launched Neeru-Meeru programme with a view to converge efforts of all the departments involved in water conservation and management and to ensure large scale participation of people.

The programme was initiated on 01.05.2000 coinciding with 12th round of Janmabhoomi programme. Neeru-Meeru activities taken up by different departments are aimed at creating more filling space for harvesting rainwater which contributes to additional ground water recharge. Seven line departments (i.e., Rural Development, Forest, Minor Irrigation (I and CAD), Minor Irrigation (PR), Rural Water Supply (PR), Municipal Administration and Endowments) are actively involved in conservation and management of water resources.

**Watersheds Development Programme Performance and Impact**

The performance of watershed programme in Andhra Pradesh is as follows:

## SCHEME-WISE WATERSHEDS

(FROM 1995-96 ONWARDS) UPTO DECEMBER 2002

| *S.No.* | *Scheme* | *No. Water Sheds* | *Area under Treatment in Lakh ha.* |
|---|---|---|---|
| 1. | DPAP | 2966 | 14.93 |
| 2. | DDP | 552 | 2.76 |
| 3. | IWDP | 842 | 4.78 |
| 4. | EAS | 1884 | 9.52 |
| 5. | APHM | 100 | 0.50 |
| 6. | RIDF VI | 1244 | 2.63 |
| 7. | RIDF VIII | 290 | 1.48 |
| 8. | WDF | 25 | 0.25 |
| | **Total** | **7903** | **36.85** |

| | | |
|---|---|---|
| Watershed projects under implementation | : | 7903 |
| Land under watershed treatment | : | 36.85 lakh ha. |
| Amount invested on participatory watershed programme (upto December 2002) | : | Rs. 992.29 crores |
| Amount generated as people's contribution for watershed development fund | : | Rs. 49.61 crores |

**Impact**

Andhra Pradesh Stale Remote Sensing Application Centre (APSRAC) is using satellite application and the satellite imageries are also used for evaluation. The Annual evaluation of watersheds in September 2002 has revealed the following impact and it shows that the programme is immensely useful to the farmers and the poor in dry areas.

| | | | |
|---|---|---|---|
| 1. | No. of districts covered | : | 20 |
| 2. | No. of watersheds evaluated | : | 5298 |
| 3. | Average increase in water levels (metres) | : | 1.96 |

| | | | |
|---|---|---|---|
| 4. | Per cent No. of wells rejuvenated | : | 43% |
| 5. | Additional area brought under cultivation (Ac.) | : | 3,34,755 |
| 6. | Decrease in labour migration | : | 61% |
| 7. | Increase in milk production (ltrs. per day) | : | 3,71,328 |
| 8. | Additional area brought under horticulture/afforestation(Ac) | : | 3,56,046 |

**D.P.A.P. KURNOOL**
**WATERSHED DEVELOPMENT PROGRAMME**

| | | | |
|---|---|---|---|
| 1. | Total Geographical Area | 17.63 | lakhs ha. |
| 2. | No. of Mandals | 53 | |
| 3. | No. of Mandals Covered | 48 | |
| 4. | No. of Habitations Covered | 436 | |
| 5. | Total Area Included | 4.40 | lakh ha. |
| 6. | No. of Micro Watersheds of 500 Ha. each | 841 | EAS-355 |
| | | | IWDP-13 |
| | | | DPAP-373 |
| | | | APRLP-100 |
| 7. | No. of Project Implementing Agencies | 71 | |
| 8. | No. of Govt. PIAs | 48 | |
| 9. | No. of Non Govt. PIAs | 23 | |
| 10. | No. of Watersheds Registered | 841 | |
| 11. | No. of Watersheds Bank Accts opened | 841 | |
| 12. | No. of User Groups | 6071 | |
| 13. | No. of Self Help Groups | 4095 | |

**D.P.A.P., KURNOOL**
**PHYSICAL AND FINANCIAL PERFORMANCE FOR 2002-03**

*Soil and Moisture Conservation*

| *Sl. No.* | *Name of the Scheme* | *Annual Target* | | *Target up to December '02* | | *Achievement up to December '02* | | *Balance* | |
|---|---|---|---|---|---|---|---|---|---|
| | | *Phy.* | *Fin.* | *Phy.* | *Fin* | *Phy.* | *Fin.* | *Phy.* | *Fin.* |
| 1. | DPAP | 3053 | 67.863 | 2291 | 39.683 | 2048 | 28.372 | 762 | 28.181 |
| 2. | EAS | 5114 | 113.687 | 3838 | 66.477 | 3317 | 45.938 | 1276 | 47.209 |
| 3. | RIDF-VI | | | | | | | | |
| 4. | RIDF-VIII | | | | | | | | |
| | **Total** | **8167** | **181.550** | **6129** | **106.160** | **5365** | **74.310** | **2038** | **75.390** |

*(Contd...)*

**Contd...**

| | | Water Harvesting Structures | | | | | | | |
|---|---|---|---|---|---|---|---|---|---|
| *Sl. No.* | *Name of the Scheme* | *Annual Target* | | *Target up to December '02* | | *Achievement up to December '02* | | *Balance* | |
| | | *Phy.* | *Fin.* | *Phy.* | *Fin* | *Phy.* | *Fin.* | *Phy.* | *Fin.* |
| 1. | DPAP | 5310 | 457.154 | 3983 | 368.335 | 11349 | 442.010 | 1327 | 88.819 |
| 2. | EAS | 7315 | 286.926 | 5255 | 283.225 | 16233 | 360.558 | 2060 | 27.521 |
| 3. | RIDF-VI | 568 | 425.780 | 406 | 280.690 | 1012 | 304.510 | 162 | 121.270 |
| 4. | RIDF-VIII | 1012 | 53.130 | 1012 | 53.130 | 1132 | 50.622 | 0 | 0.000 |
| | **Total** | **14205** | **1222.990** | **10656** | **985.380** | **29726** | **1157.700** | **3549** | **237.610** |

*(Contd...)*

**Contd...**

| Sl. No. | Name of the Scheme | *Afforestation and Horticulture* | | | | | | | |
|---|---|---|---|---|---|---|---|---|---|
| | | *Annual Target* | | *Target up to December '02* | | *Achievement up to December '02* | | *Balance* | |
| | | *Phy.* | *Fin.* | *Phy.* | *Fin.* | *Phy.* | *Fin.* | *Phy.* | *Fin.* |
| 1. | DPAP | 3686 | 197.108 | 3686 | 118.965 | 4995 | 96.025 | 0 | 78.143 |
| 2. | EAS | 5725 | 272.351 | 5725 | 141.443 | 7639 | 97.631 | 0 | 130.907 |
| 3. | RIDF-VI | 450 | 57.851 | 450 | 57.851 | 450 | 57.851 | 0 | 0.000 |
| 4. | RIDF-VIII | | | | | | | | |
| | **Total** | **9861** | **527.310** | **9861** | **318.259** | **13084** | **251.507** | **0** | **209.050** |

*(Contd...)*

**Contd…**

| Sl. No. | Name of the Scheme | *Trainings, Comm. Org. and Est.* | | | | *Total* | | | |
|---|---|---|---|---|---|---|---|---|---|
| | | *Annual Target* | *Target. up to December '02* | *Achievement up to December '02* | *Balance* | *Annual Target* | *Target up to December '02* | *Achievement up to December '02* | *Balance* |
| 1. | DPAP | 114.794 | 86.094 | 62.951 | 28.700 | 836.919 | 613.077 | 629.358 | 223.843 |
| 2. | EAS | 192.306 | 144.226 | 101.926 | 48.080 | 865.270 | 635.371 | 606.053 | 253.717 |
| 3. | RIDF-VI | | | | | 483.631 | 338.541 | 362.361 | 121.270 |
| 4. | RIDF-VIII | | | | | 53.130 | 53.130 | 50.622 | 0.000 |
| | **Total** | **307.100** | **230.320** | **164.877** | **76.780** | **2238.950** | **1640.119** | **1648.394** | **598.830** |

## WATERSHED DEVELOPMENT PROGRAMME: KURNOOL DISTRICT

| | | |
|---|---|---|
| | Total Watersheds | 841 |
| *(a)* | Completed | 260 |
| *(b)* | Ongoing | 581 |
| | Mandals covered | 48 |
| | Villages covered | 436 |
| | Area covered (lakh ha.) | 4.40 |

19

## The Challenges of Globalisation

Globalisation gives rise in some quarters to fears that can lead to suspicion, protectionism, and policies that are ultimately self-destructive. Such fears cannot be allowed to frustrate the great potential of a world in which countries drawing closer together. We believe that countries can face the challenges of globalisation positively, demanding as those challenges may be.

All countries can benefit from full participation in the world's markets, including its financial markets. Protectionist pressures must be resisted and reversed, and the principles of openness and mulilaterism promoted by the World Trade Organisation, the IMF, and the World Bank must be honoured. And financial market integration should be seen as a positive force: it offers access of global financial intermediation and a stimulus for more competitive and efficient domestic financial sectors; and it promotes efficiency and growth worldwide.

How encouraging it is, therefore, to see that so many developing countries in transition have been freeing up their trade and exchange systems within the framework of our structural adjustment programmes.

No country can afford to forgo the benefits of integration into global market: the alternative is marginalisation and stagnation. But all countries must take the steps to minimize the associated risks. More than ever before, countries need tightly disciplined macro economic policies to maintain a stable environment for investors, whether domestic or foreign. And while foreign capital can be a useful—and sometimes vital complement to domestic saving, it is not a substitute for it: domestic saving remains the key to investment and sustainable growth. It is also clear that strong financial institutions are essential to avoid market disturbances at home and to secure an effective defence against external pressures. Competitive banking and financial systems that are sound, well regulated, and properly supervised are indispensable for countries to be able to expose their economies safely to the pressures that can arise in global markets.

The challenges to globalisation therefore add to the need for the developing and transition countries to press ahead with their adjustment and reform efforts. For many, this means creating conditions to attract foreign financing and use it effectively. But a growing number of countries have been facing a different problem: how to cope with large-scale capital inflows. Such inflows, especially when they are easily reversible; provide no grounds for relaxation of adjustment and reform.

They should not be used to finance domestic consumption. In many cases, they call for stronger fiscal discipline; and in some cases, exchange rates should be allowed to take part of the strain. Many developing countries and countries in transition also, of course, need to do more to deepen and widen the role of market forces and to foster more competitive market environments in order to promote transparent and efficient mechanisms for resource allocation.

Is globalisation any less demanding for the industrial countries? Not at all! It adds to the urgency of the task of taking full advantage of the current expansion to tackle the deep-rooted problems that are limiting the pace, the quality, and perhaps, the sustainability of their growth.

All has to applaud the increased efforts and commitments to reduce fiscal deficits, but in most cases underlying imbalances remain large and the pace of consolidation too slow. More must be done not only to redress present imbalances but also to meet the growing demands of the future.

Another deep-rooted problem—structural—unemployment must also be tackled sooner, rather than later. Budget laxity and high unemployment tend to feed on each other. While cyclical conditions provide the opportunity, governments must not flinch from the task of improving the functioning of labor markets. How? It is not an easy task: by reforming regulations and policies that impede employment creation and job search.

Monetary stability, macroeconomic discipline, sound financial systems, and efficiently working market mechanisms are essential for all countries that embrace globalisation. But they are not sufficient for any. To fight the fears that globalisation sometimes inspires, countries need policies that promote not just economic efficiency, financial stability, and growth but also equity and high quality growth. In too many countries, the quality of growth suffers from widening distributional inequalities related partly to high unemployment but also stagnating wages of unskilled workers. And too many countries continue to suffer from poor governance, corruption and increasing crime.

Of course, economic policy can provide only part of what is needed to rid the world of these blights. But it is a vital part. To promote equity, efficiency, and sustainable growth, governments carry an inescapable responsibility for investment in human capital—through education, health care, and well-targeted social safety nets—and also for establishing and maintaining honest and effective systems of public administration, law, order and justice. If these essential services are to be affordable, there is certainly no room for unproductive expenditures—military or otherwise—and wasteful subsidies: they must bear the brunt of fiscal consolidation. So globalisation demands a lot from governments if it is to deliver its promise of stronger and high-quality growth.

## REFERENCES

A.K. Sen: "Pattern of British Enterprises in India: 1854-1914 in Social and Economic Development, Singh and V.B. Singh (eds), New Delhi, 1965, p. 420.

Cottrel, P.L.: British Overseas Investment in Nineteenth Century, Macmillan, 1975.

Murphy, Rhoads: The Outsiders: Western Experience in India and China (University of Michigan Press, 1977).

Das, Parekh and Parekh: India Development Report (1999), Oxford University Press.

Reich, Robert (ed): The Power of Public Ideas, Ballinger, Cambridge, Mass USA, 1988.

## The Truth About Global Competition

### *The Economic Myths Behind Globalisation*

Local communities everywhere are on the front lines of what might well be characterized as World War III. It is not the nuclear confrontation between East and West—between the Soviet Union and the United States—that we once feared. It is a very different kind of conflict. There is no clash of competing military forces and the struggle is not defined by national borders. But it does involve an often-violent struggle for control of physical resources and territory that is destroying lives and communities at every hand. It is a struggle between the forces and institutions of economic globalisation and the communities that are trying to reclaim control of their economic lives. It is a conflict between competing goals—economic growths to maximize profits for absentee owners versus creating healthy communities that are good places for people to live. It is a competition for the control of markets and resources between global corporations and financial markets on the one hand and locally owned businesses serving local markets on the other.

Two things of fundamental importance to each and every one of us are now very much at stake.

- Will people and communities control their local resources and economies and be able to set their own goals and priorities based on their own values and aspiration? Or will these decisions be left to global financial markets and corporations that are blind to all values save one—instant financial returns?

- Will the life sustaining resources produced by the regenerative capacities of our planet's ecosystems be equitably shared to provide for the material needs of all of us who inhabit this bountiful planet, as well as for our children and their children unto the seventh generation and beyond? Or will we allow a global economic system that is now functioning on auto-pilot beyond conscious human control to consume and destroy the ecosystem and our social fabric in its insatiable quest for money?

Economists, politicians, corporate spokespersons and the media have for years been touting the benefits of the global economy. They have called on us to support trade agreements such as the North American Free Trade Agreement (NAFTA) and the World Trade Organisation (WTO) to remove the constraints of economic borders and open to everyone the opportunities of growth and prosperity in the global economy. They have promised rich rewards for those workers and communities that become successful global competitors.

Many of the most ardent boosters of economic globalisation met earlier in the year at the annual meeting of the World Economic Forum. This Forum has for years brought together top industrialists and political figures from around the world to advance the proposition that removing tariffs and other restrictions on the free international flow of trade and money is a key to creating new economic opportunity and prosperity. It thus caused quite a stir when the Forum publicly announced that economic globalisation is producing disastrous consequences that threaten the political stability of the Western democracies. Their warning bears close examination for being one of the most honest and accurate assessments of the consequences of economic globalisation yet produced by leading advocates of that process. The observation is that:

- Economic globalisation is causing severe economic dislocation and social instability.
- The technological changes of the past few years have eliminated more jobs than they have created.

- The global competition "that is part and parcel of globalisation leads to winner-take-all situations: those who come out on top win big, and the losers lose even bigger."
- Higher profits no longer mean more job security and better wages. "Globalisation tends to delink the fate of the corporation from the fate of its employees."
- Unless serious corrective action is taken soon, the backlash could destabilize the Western democracies.

We don't have to go far to find examples of what they are talking about and why people are getting a bit upset as they wake up to the realities of who is winning in the ruthless competition of the global economy. The disparities between the winners and losers in the global competition are becoming more obscene with each passing day.

We are coming to realize that the extravagant promises of the advocates of the global economy are based on a number of myths that have become so deeply embedded in Western industrial culture that we have grown to accept them without examination.

- The myth that growth in GNP is a valid measure of human well being and progress.
- The myth that free unregulated markets efficiently allocate a society's resources.
- The myth that growth in trade benefits ordinary people
- The myth that global corporations are benevolent institutions that if freed from governmental interference will provide a clean environment for all and good jobs the poor.
- The myth that absentee investors create local prosperity.

**The Growth Myth**

Our measures of growth are deeply flawed in that they are purely measures of activity in the monetized economy.

Expanded use of cigarettes and alcohol increases economic output both as a direct consequence of their consumption and because of the related increase in health care needs. The need to clean up oil spills generates economic activity. Gun sales to minors generate economic activity. A divorce generates both lawyers fees and the need to buy or rent and outfit a new home increasing real estate brokerage fees and retail sales. It is now well documented that in number of other countries the quality of living of ordinary people has been declining as aggregate economic output increases.

The growth myth has another serious flaw. Since 1950, the world's economic output has increased 5 to 7 times. That growth has already increased the human burden on the planet's regenerative systems—its soils, air, water, fisheries, and forestry systems—beyond what the planet can sustain. Continuing to press for economic growth beyond the planet's sustainable limits does two things. It accelerates the rate of breakdown of the earth's regenerative systems—as we see so dramatically demonstrated in the case of many ocean fisheries, and it intensifies the competition between rich and poor for the resource base that remains.

This is vividly illustrated by many of the development projects in India many funded with loans from the World Bank and other multilateral development banks—that displace the poor so that the lands and waters on which they depend for their livelihood can be converted to uses that generate higher economic returns—meaning converted to use by people who can pay more than those who are displaced.

**The Myth of Free Unregulated Markets**

It is almost inherent in the nature of markets that their efficient function depends on the presence of a strong government to set a framework of rules for their operation. We know that free markets create monopolies, which government must break up to maintain the conditions of competition on which market function depends.

We also know that markets only allocate efficiently when prices reflect the full and true costs of production. Yet in the absence of governmental regulation, market incentives persistently push firms to cut corners on safety, pay workers less than a living wages, and dump untreated toxic discharges into a convenient river. In our present competitive context if management does not take such measures, they are likely to be replaced by the owners or bought out by someone with less scruples who will.

**The Myth of Free Trade**

Many so-called trade agreements, such as the North American Free Trade Agreement (NAFTA) and the World Trade Organisation (WTO) are not really trade agreements at all. They are economic integration agreements intended to guarantee the rights of global corporations to move both goods and investments wherever they wish—free from public interference and accountability. WTO is best described as a bill of rights for global corporations.

**The Myth that Economic Globalisation is Inevitable**

Many of the people who claim globalisation is a consequence of inevitable historical forces are paid to promote that message by the same global corporations that have invested millions of dollars in advancing the globalisation policy agenda.

**The Myth that Corporations are Benevolent Institutions**

The corporation is an institutional invention specifically and internationally created to concentrate control over economic resources while shielding those who hold the resulting power from liability for the consequences of its use. The more national economies become integrated into a seamless global economy, the further corporate power extends beyond the reach of any state and the less accountable it becomes to any human interest or institution other than a global financial system that is now best described as a gigantic legal gambling casino.

All over the world people are indeed waking up to the truth about economic globalisation and are taking steps to reclaim and rebuild their local economies. Such communities face basic choices as to how they will divide their efforts between competing for a share of the declining pool of good jobs that global corporations offer and working to create locally owned enterprises that sustainably harvest and process local resources to produce the jobs and the goods and services that local people need to live healthy, happy, and fulfilling lives in balance with the environment.

Our experience with the real consequence of economic globalisation is pointing to many important lessons. One such lesson is that economies should be local, rooting power in the people and communities who realize their well-being depends on the health and vitality of their local ecosystem. If it is protectionist to favour local firms and workers who pay local taxes, live by local rules, respect and nurture the local ecosystems, compete fairly in local markets, and contribute to community life—then let us all proudly proclaim ourselves to be protectionist.

Such choices are not isolationist, to the contrary, they create a foundation for creative cooperation with our neighbours—whether they be in the United States or in other countries—to share experience, ideas, and technology—and to join in international solidarity in rewriting the rules of the global economy to favour local over global businesses, and to encourage cooperative relations among people and communities. It is our consciousness—our ways of thinking and our sense of membership in a larger community—not our economies—that should be global.

Millions of people are also making an important discovery—that life of is about living not consuming. A life of material sufficiency can be filled with social, cultural, intellectual, and spiritual abundance that place no burden on the planet.

It is time to assume responsibility for creating a new human future of just and sustainable communities freed from the myth that greed, competition, and mindless consumption are paths to individual and collective fulfilment. It will take millions of people around the world linked together into a powerful political coalition aimed at radical, political and economic reform to win the war that global capital is waging against us.

21

## Globalisation: *A Moral Imperative*

Globalisation has become today's buzzword. It has also become a battle ground for two radically opposed groups. There are the anti-globalists, who fear globalisation and stress only its downside, seeking therefore powerful interventions aimed at taming, if not (unwittingly) crippling it. Then there are the "globalist" (a class to which I belong) who celebrate globalisation instead, emphasize its upside, while seeking only to ensure that its few rough edges be handled through appropriate policies that serve to make globalisation yet more attractive.

Many anti-globalists consider the central problem of globalisation to be its amorality, or even its immorality. But these critics have too blanket an approach to globalisation. The word covers a variety of phenomena that characterize an integrating world economy: trade, short-term capital flows, direct foreign investment, immigration, cultural convergence et al. The sins of one of the above cannot be visited upon the virtues of another. Some are benign even when largely unregulated whereas others can be fatal if left wholly to the marketplace.

In particular, the freeing of trade is largely benign: if I exchange some of my toothpaste for some of your toothbrushes, we will both be better off than if we did not trade at all. It would require a wild imagination, and a deranged mind, to think that such freeing of trade leads to debilitating economic crises. Equally, it is illogical to believe, as non-economists who fear globalisation do, that freeing of trade is bad because the freeing of short-term capital flows led to a debilitating financial and economic crises and could do so again. In fact, while there are some obvious simulates between free trade and free capital flows, e.g. that segmentation of markets creates efficiency losses—the economic and political dissimilarities are even more compelling and policy makers cannot ignore them.

Anti-globalist critics are in fact often reacting viscerally to a much larger issue: the victory of capitalism over its arch rival, communism. For campus idealists who have always looked for alternatives to what they conventionally consider to be the greed and lack of social conscience that characterize capitalism, the situation is psychologically intolerable Some have turned to street theatre, nihilism and the anti-intellectualism that has been manifest in the last few years. The more sophisticated have succumbed to a stereotypical representation of corporation as the evil forces of capitalism that have captured the state, democratic institutions, and even international bodies such as the World Trade Organisation.

What these critics often forget is that certain economic freedoms are basic to prosperity and social well-being under any conditions, and are thus of the highest moral value. Property rights and markets, for instance, provide incentives to produce and allocate resources efficiently, and can in turn strengthen democracy by allowing a means of sustenance out side pervasive government structures. The quality and breadth of democracy can then be enlarged as excluded groups, such as women and the poor, are pulled into literacy, gainful employment and better health through higher public spending or the spread of economic incentives.

Critics nevertheless go on to maintain that the global spread of free markets and free trade is responsible for continuing poverty in poor countries, and for alleged growth in inequality between and within countries. Labour unions in the rich countries also fear that trade in cheap labour-using goods from poor countries.

But I do not think these concerns are well-founded. In India which has almost a quarter of the world's poor, there is good evidence that autarchic and anti-market policies produced abysmally low growth rates at 3.5 per cent annually over a quarter of a century, with a correspondingly negligible impact on poverty has declined. Higher growth rates in turn depend on several factors; but openness to trade and direct investment and a skilful use of markets are definitely and important contributory factor.

As for inequality among nations, it is precisely those countries that embraced integration into the world economy, i.e. the Far Eastern Four and then the ASEAN countries, which raced ahead with dramatic growth rates whereas several countries of Africa, Latin America and Asia that looked inwards failed to deliver growth and also made little dent on poverty.

The evidence on trade and investment impoverishing our workers is also flawed. My own research suggests that the downward pressure on workers' wages due to technical change has been dampened, not magnified, by trade with the poor countries. Research also shows that big corporations use abroad technologies similar to those at home, instead of exploiting lower standards or forcing them yet lower through their financial clout.

One result of these mistaken arguments against globalisation has been an insistent clamour for certain environmental and labour standard to be linked to rules on international trade. But by seeking to create new obstacles to free trade, you undermine the freeing of trade, while mixing up trade with a moral agenda undermines that very moral agenda. It gives other countries the definite impression that you are using ethical rhetoric to mask protectionist self-interest.

The notion that global free trade and investment are responsible for poverty, inequality, lowering of standards and harming social progress is little short of astonishing. Yet national politicians and international bureaucrats give it who think that going along is way of getting along. In denying the virtues of globalisation, they actually harm the very causes they profess to embrace.

## High World Trade Growth *Vs.* Output
*WTO Sees Link to Globalisation*

World trade in merchandise goods is expected to increase in volume by 8 per cent in 1995 down marginally on the very high 9½ per cent for 1994. Although the current outlook is for a further modest slowing next year, trade growth will remain above the average of the past decade.

Recent trade growth figures continue to exceed world production growth by a large margin in 1995 probably by a factor of almost three and next year close to double. This persistent pattern relates closely to the "globalization" of the world economy; a process which, brings far-reaching benefits and which can be promoted through the further development of the multilateral trading system.

The recent growth is as follows:

- a 13 per cent rise pushed the value of world merchandise trade past the $ 4,000 billion mark for the first time, to $ 4,090 billion.
- an 8 per cent increase in the value of trade in commercial services, to $ 1,100 billion, after near stagnation in 1993;
- a 23 per cent increase in the dollar value of merchandise trade in the first six months of 1995 which, allowing for the depreciation of the US dollar, is consistent with a full-year growth in trade volume of 8 per cent.

## Globalisation

Over the period from 1950 (when the process of trade liberalisation through the early GATT Round got underway) to 1994, the volume of world merchandise trade increased at an annual rate of slightly more than 6 per cent and world output by close to 4 per cent. Thus, during those 45 years wörld merchandise trade multiplied 14 times and output 5½ times. However, the excess of trade growth over output growth varied; from an average of a mere half percentage point in the period 1974-84 to nearly 3½ percentage points in the most recent 10 years. In fact, the excess during the years since 1990 has been much higher still but it is not yet clear whether or not this represents a permanent shift to a faster rate of increase in the world's trade-to-output ratio.

To the question "will globalisation continue?" In this regard one has to observe two factors—technological change and the evolving strategies of firms and individual investors—impart a natural momentum to global integration. It is government policies which can speed-up, slow down or even reverse progress on global integration. In this context, the role of non-discrimination—in particular, through the "most-favoured-nation" (MFN) clause—is examined.

## The MFN Clause

MFN was the centrepiece of a multiplicity of bilateral trade agreements reached in Europe in the second half of the 19th century, a period marked by very low tariffs and rapidly increasing trade. In contrast, the 1920s and the 1930s saw effects to restore liberal trade through international trade conferences rather than legally-binding commercial treaties based on MFN. The failure of these efforts contributed to the Great Depression and provided some of the roots of military confrontation in 1939. It was only after the War that negotiations established what became the GATT, a multilateral contract consisting of rules and disciplines and based firmly (Article) on MFN treatment.

The GATT system has been a post-war bulwark against a return to the trade chaos of the 1930's. In the 1990's, a disintegration of the globalised international economy on the scale of 1930's is almost unthinkable. In contrast, today "the threat that would be posed by a loss of credibility of the multilateral rules" (now represented by the WTO) would be "a fracturing of the global economy into inward-looking and potentially antagonistic trading blocks".

One can suggest two safeguards against such an eventuality:

- the examination of new ways to ensure that free-trade areas and customs unions remain outward-looking and complement rather than compete with the multilateral trading system; and
- progress in dealing, at the multilateral level, with new issues tied directly to the further evolution of the global economy. These include telecommunications, financial services, environment, competition and investment policies among others.

Progress in dealing with these and other issues at the multilateral level will have a significant impact on the future pace of global integration, both directly and through its impact on the credibility of the multilateral system in influencing the broad spectrum of national trade policies.

23

## Urbanisation and Globalisation

How we handle globalisation will determine whether our cities and our civilisation will be divided and violent or user-friendly and peaceful. We cannot get a clear picture of urban life in the 21$^{st}$ century, especially in the poor countries of the South, unless we take into account the phenomenon of globalisation, which has already brought dramatic changes make their first appearance. So it is there too that the great upheavals of the next century will take place.

Globalisation gives shape to the "Global Village". The "information era" that it ushers in compress time and we are now living in a world speeded up as never before. Worldwide urbanisation is proceeding at a similar rate and its pace in the poor countries of the South seems terrifying. By 2025, two-thirds of humanity will be living in cities and towns, where the best opportunities in life tend to be.

Globalisation also accentuates a "new urban geography" in both North and South. Islands of rich consumers are springing up in cities amid an ocean of deprived people. More and more unemployed people, immigrants, minorities and the homeless, are pushed into cities by pressure from "market economies". As a result, all urban area—not just those in the poor countries of the South—will have to deal with growing internal tensions. In New York, for example, the poorest 20 per cent of the population earns 15 times less than the richest 20 per cent.

Cities have always had their smart neighbourhoods and their dangerous areas. But such social and geographical segregation has changed in pace and scale because of the growth in the urban population, the increase in "illegal" migrant and rising uncertainty.

In fact, we have entered a period of historical transition, where discontinuities prevail over adjustment. Radical changes in the nature of production and jobs and the incredible concentration of capital in the hands of the financial sector and speculators weigh much heavier in our lives these days than the state's efforts to adjust and improve the market economy. Segregation in cities has been given a new lease of life whose consequences we do not know. It has reached unprecedented dimensions because of the explosive growth of urban areas.

***According to one scenario, things will go badly.*** The growing pace of globalisation will increase uncertainty about the future. Fear and defence mechanisms will grow among people and institutions, fuelling intolerance, xenophobia and mistrust of everything new or foreign. Urban tensions will manifest themselves with increasing violence, and segregation will sharpen. Public areas will be abandoned and become dangerous no-man's lands, the wretched abode of society's rejects. Cities will lose their original function of being a crossroads for meeting and exchange.

If globalisation also continues to go hand in hand with deregulation of financial markets and an unchanged level of indebtedness of poor countries, the latter will not be able to maintain their urban infrastructures. And if on top of this there is corruption and lack of political will, challenges to the system will increase and violence will grow. Cash-strapped authorities will respond with undemocratic mafias which provide them with funds.

***According to a second scenario, everything will be all right.*** In line with the principle that "everything the state does is public, but the state doesn't control everything that is public," a new social contract will be drawn up between the state, the market, the working population and civil society, including NGOs. Cities will develop a new quality of life by providing citizens with forum for exchange. Jobs will be created in the social sector, in the fields of the environment, education, research, culture and leisure, opening up possibilities for young people.

In the countries of the South, long-term development strategies will be drafted and urban planning practiced, taking advantage of the opportunities provided by globalisation but without falling into its traps. Town planning will become part of the political process, and the state will work with the private sector, monitored by institutions of civil society. Adequate housing will be built with the help of micro-credit and controls on the price of building materials. Improved infrastructures will enable marginal areas to become part of the civilized part of the city. Democracy will come up with new ways of governing with the help of networks of involved citizens.

***In a transitional scenario,*** action strategies should fall somewhere between these two extremes. They should include social goals so that in big urban areas a society emerges which is founded on participatory democracy and on "capitalism with a human face" or "market socialism".

But the outlook is less clear than ever. Let us hope the present transition will lead rapidly to a new revival of humanism, whose first signs we are already seeing. This would open up the road to a development which is fair, humane and peaceful.

24

## Myths and Illusions

The tide of precarity is rising steadily, so that people who have never been poor no longer regard poverty as a distant prospect but as one so close that it could engulf them at any moment.

In 1989, the fall of the Berlin Wall was rightly welcomed because it marked the collapse of a system that provided a degree of equality but rejected freedom. Today there is a strong possibility that the system gradually spreading all over the world—a kind of neo-liberal fundamentalism—may also collapse. In its obsession with freedom, vital though freedom is, this fundamentalism disregards equality, a term which should not be regarded here in purely static and statistical terms, but as something dynamic and ethical. Equality can only be truly practised in a context of social solidarity or to borrow from the vocabulary of the French Revolution of fraternity.

On the one hand, we have a world that is immensely rich in resources, possibilities, knowledge and experience; its constituent societies are freer and more dynamic than ever. There is an extraordinary potential for everyone to live a better life. But at the same time, new and ever higher walls are being built both between peoples and between social groups within individual countries. We are experiencing a travesty of development, which is creating a world bipolarized into extremes of wealth and poverty.

The most common reactions to this disastrous situation are very often the result of two misapprehensions. The first can only be described as ideological or doctrinaire since it is not based on the facts as they can be observed. It says that since the dominant system of values and things is by definition more than satisfactory, the persistence of impoverishment is merely a temporary blip. Enough time has elapsed, however, for us to see that this is not the case, including in countries where this system has been part of the established order for more than a century. One statistic is particularly eloquent. In just over 30 years, world production has approximately doubled, but the gap has more than doubled between the income of the 20 per cent of world's people living in the richest countries and the income of the world's poorest 20 per cent, according to the United Nations Development Programmes.

The second misapprehension stems from another form of blindness and illusion, namely the belief that poverty can be regarded exclusively as a moral issue, as if it had no other kind of implications for those who are not poor. Globalization is, however, a two-way process. It enable the countries of the North to export their values and their paradigms as well as their goods and capital to the countries of the South, but it also makes them much more vulnerable to the backlash of crises that afflict these countries. Even in the North, the cult of competitiveness is undermining situations once considered extremely stable. The tide of precarity is rising steadily, so that people who have never been poor no longer regard poverty as a distant prospect but as one so close that it could engulf them at any moment.

Because of inadequate socio-economic development, the extraordinary upsurge of democracy over the past 30 years remains a very fragile process, and there is a risk that the trend may be reversed. When hunger, disease and ignorance prevail, citizens' participation in decision-making becomes either non-existent or a mere charade. Democratic institutions become empty shells, representational bodies existing in form only and devoid of real significance.

Social divisions caused by economic distortions exacerbate the failures of democracy which in turn pose serious threats to civil order within countries and to peace between nations. It is high time to face these obvious facts.

25

# Globalisation and Knowledge Divide

Globalisation looks very different when it is seen, not from the capitals of the West, but from the cities and villages of the South, where most of humanity lives. Four examples taken from India, illustrate how the paradoxical forces shaping globalisation look when seen from the other side.

Five school children died in a remote village in India after drinking water and powdered milk mixed in a vat that had contained a powerful insecticide. Nobody could read the label of the vat and the children were poisoned. The insecticide in question has been banned in practically every industrialized nation; its sale continues only in places like my country.

Secondly, an important annual event recently took place in North India. Potato growers gather there to exchange the best seeds they have produced in the last year. It is an act of pride for communities to share with others seeds that will help improve the production of potatoes. A transnational corporation attended the festival and are now working to patent the genes of these traditional foodstuffs in order to sell them as profit.

India's macro-economic indicators are excellent. In the offices of investment bankers, you will be told that India is a great investment opportunity. The situation is not so rosy, however. Thirty per cent of the population have been living below the poverty line for the last so many years. Ten per cent of the population are living below the critical poverty line: their income is insufficient to pay for even minimal nourishment. So much of the workforce is unemployed or under-employed.

A distinguished North American political scientist, Dr. Benjamin Barber, recently pointed out that in the United States democracy had degenerated into bringing one group of rascals in for four years, and then throwing them out and replacing them with another group of rascals for four years. From the perspective of the South, that looks very good! In a context where rascals manipulate elections and stay in power for fifteen or sixteen years, I would appreciate the chance to throw them out through peaceful elections every four years.

Thus, the complaints of the North are often the aspirations of the South. Progress in industrialized nations can be a threat to developing countries.

Ten year ago, in the euphoria of globalisation and the expansion of services and finance that followed the fall of the Berlin Wall, I advanced the idea that we were entering a fractured global order. Globalisation brings us into contact with one another, but it also strengthens profound divisions and fractures in terms of societies and income, and most importantly in our capacity to generate and utilize knowledge. Over the last ten years, the concentration of wealth and power has greatly increased both within and between societies.

There is a real risk of two civilisations emerging, with two ways of viewing and relating to the world: one based on the capacity to generate and utilize knowledge, the other passively receiving knowledge from abroad and deprived of the ability to modify it.

The world now faces the prospect of this Knowledge Divide becoming an unbridgeable abyss. We need the international community to return to the basic principles of international co-operation and introduce the idea that a minimum level of science and technological capability, including access to the Internet, is an absolute necessity for developing countries and should be the subject of international solidarity.

This can be achieved. However, contrary to the situation of 20 years ago, national governments are no longer the major players in the game of science and technology. Whether we like it or not, the private sector and the international community of scholars must be invited to the table with governments from the North and South to begin discussing an agenda for the mobilisation of a science and technology for development. United Nations with a mandate for the development of the sciences, has a special role to play in the revitalisation of international co-operation in this field.

26

## The Nation-State and Globalisation

The world has changed dramatically. Some of the changes are as yet only dimly understood. We are all going to be confronted with many challenges to the whole concept of government and to the role of the nation-state as we move into the next century.

There are two principal aspects to these changes. Globalisation of the world economies is sharply limiting the independence of action of the nation-state. In addition, we are only just beginning to understand what the existence of one superpower, supreme militarily, financially, means to the evolution of world diplomacy and world polities.

These remarks are directed to the first aspect. Governments are now losing influence. Private enterprise, capitalism, summarized as 'the market', is gaining power. Privatisation is a keyword. Across the political spectrum, liberal conservative and formerly socialist parities have all accepted the downsizing of government, the privatisation of many activities and the reduction of government debt. Governments in crisis in the developed or in the developing world have been left in no doubt about what they should do.

The International Monetary Fund and the World Bank have made it clear that assistance would not be available to countries in distress unless appropriate policies were put in place, and IMF prescriptions often involve substantial and detailed microeconomic reform within a country with considerable hardship for its people.

Meanwhile, competition for international capital has become much more severe. In the early independence years, Commonwealth countries believed they could write their own internal rules about the performance and behaviour of capital. Now those rules have to be rewritten to maximize international attraction. The relationship has to be competitive; the rules have to be friendly to capital. This is a totally different environment from the one in which most Commonwealth countries gained their independence in the immediate post-war years.

The new global organisation of industry has significant consequences for social policy. Many governments would have conducted policies designed to see that workers gained a fair share of the returns of an enterprise. With the globalisation of industry, such policies are no longer possible. Governments now tend to argue for lower wages, for smaller workforces, to maximize the competitiveness of their particular country as a home for global corporations. This has consequences of enhancing the profit share as opposed to the wage share of a particular enterprise.

One direct consequence of these changes is a significantly growing disparity in wealth between rich and poor in all countries worldwide. This may not matter so much if the poor were also becoming better off compared to their own earlier standards but in many cases this is not so. The idea of a living wage is no longer relevant. Workers in some countries are often paid a wage which could not support even the smallest of families. In this day, if that is what the market determines, then that is what must happen.

In today's world, governments must fashion their policies to meet the wishes of the international market place. There are fundamental differences from earlier times. The global organisation of industry in which national boundaries become irrelevant is certainly new. Some aspects of information technology can operate much faster and with more devastating effect than the old cable system of the last hundred years. This has led to an explosive growth in financial markets. The volume of money traded each day is huge (and) through modern communications, this finance has great mobility.

We all know enough of markets to know that they favour the powerful, the united and the strong and that markets can overwhelm and destroy smaller players. Sometimes smaller players are entire nations.

Those who suggest that the markets alone must be allowed to determine economic outcomes favour a world in which the large will do much better than the small. So far as countries are concerned, most Commonwealth countries are in the smaller category in a world in which large financial institutions and manufacturing corporations operating globally will dominate trade and commerce.

For most countries, banks and financial institutions, which are part of the culture of that country, will become a matter of the past. Banking services will be American, European, Japanese or perhaps Chinese. The consequences of this market dominance are clear. Corporations need a global spread and many national rules for the good order and conduct of business and commerce will no longer be relevant.

For the world as a whole, the most serious problem is volatility, possibly leading to systematic breakdown. Since the Asian economic problems of 1997, there has been a great deal of discussion about the present system and about changes that need to be made.

For a while it appeared that the United States really was going to move the reform process forward but now the tendency seem to be 'it's all right, we have escaped, leave well enough alone' ...There is a need to reform the system, to establish much tougher international rules for prudential supervision and control. The IMF has demonstrated time and time again that it is not interested in avoiding crises, it is only interested in picking up the pieces after they have occurred. If this is its charter, it certainly needs reviewing. The IMF's present operations are inadequate.

Since governments have seemingly lost significant power to corporations and to financial markets and since they do operate within an increasingly globalised framework, individual governments are not capable of undertaking this task. The task is international and global. Whether it is a reformed IMF or a new institution is a matter for debate.

At their last meeting the Commonwealth Finance Ministers pointed to a number of changes, most of which are desirable, but there was no sense of great urgency, no sense of dynamism. They spoke of a need for new financial market architecture but nobody has tried to spell out what—means.

There are two specific tasks: how to preserve some form of equity and reasonable competition in a globalised market place and how to establish stability within the financial markets themselves.

The IMF's financial resources should be strengthened as a means of averting crises through the provision of contingency funds. Immediate access to adequate funding can be essential for this purpose if crises are to be avoided. Finding a way to encourage the IMF to help avert crises instead of just reacting to crises after they have occurred is a most important requirement.

In any liquidity arrangement, assisting a country is distress, the IMF should take care not to absolve lenders of their responsibility. In some cases IMF should take care not to absolve lenders of their responsibility. In some cases IMF bailouts have done more to help the lenders than the countries themselves. The lenders need to carry their own risk.

For poor countries, how to protect themselves and advance the welfare of their own people in an unpredictable world is a major challenge and very often a major problem. Apart from moves to establish greater stability designed to avoid systematic breakdown within the world's financial system, there also need to be urgent moves to establish an international body to establish rules for fair trading in a globalised environment. Middle ranking and small countries would have most to gain from such an innovation.

## Add Value, Go Global

### *Can Southern Firms Break into Export Markets?*

The global economy has changed beyond recognition over the last decade. Widespread economic policy reform and in particular trade liberalisation have opened up new opportunities for developing countries. In poor countries, however, the consequences of trade liberalisation are not always positive. What can the private sector do to respond better and make the most of new trading opportunities? What factors have limited the impact of economic reforms on export performance?

Why have exports from poorer countries failed to increase more rapidly following trade liberalisation? What can be done to improve performance? Research on the response of firms in the private sector to economic reform can underpin new approaches to export promotion for poorer developing countries. For a long time, protective trade policies, poorly performing state-owned industries and state controls over the private sector were blamed for poor export performance in Africa and South Asia. Now that some of these problems have been remedied, other obstacles have come to light.

The effect of economic liberalisation and adjustment on the performance of poor countries has been cause for concern. Trade liberalisation should increase incentives to export and facilitate business enterprise by encouraging private ownership through privatisation and by attracting foreign investment. Macroeconomic stability ought to boost business confidence and performance. All these factors should promote exports, offsetting job and income losses caused by the closure or reorganisation of inefficient enterprises and industries yet, although some degree of reform and stability it is without export growth that was expected.

Trade reform and macroeconomic stability may be necessary conditions for improved export performance put by them are insufficient. The obstacles to improving export performance are numerous and there is no easy policy answer. The research programme examined export performance at three levels.

- Regional: how trade strategies should vary with skills and natural resource endowments
- National: factors influencing the export performance of manufacturing
- Sectoral: the performance of particular sectors of the economy.

The East Asian economies have shown that developing countries can compete successfully in global markets. For many, they provide a blueprint for economic growth applicable to many poor countries.

South Asia's comparative advantage lies in its abundant unskilled labour, while Africa's lies in its abundant natural resources. Different export promotion strategies are essential. South Asia's best prospects are in labour-intensive manufacturing: the region's low level of exports would soar over the next decade if current obstacles to trade were reduced. Africa's exports could also increase but its biggest potential in primary products that need little educated labour and abundant natural resources.

Some African countries could also be substantial exporters of manufacturers, but their actual manufactured exports in most cases now fall far short. Comparing Ghana to Mauritius—one of Africa's most successful exporters of manufactured goods differences in firm-level efficiency are apparent Mauritian firms have more capital per worker and use it more efficiently. Reducing trade barriers is not sufficient. Wages in Ghana would have to be substantially lower to offset low labour productivity. Alternatively, labour productivity will have to be drastically improved if Ghanian firms are to compete successfully in export markets with wages at current levels.

Even when companies use capital and labour efficiently, poor infrastructure is a frequent stumbling products to export markets—an acute problem in landlocked countries and equally acute for manufacturers as research on Uganda clearly shows. What huts manufacturing exporters is being hit by the high cost of transporting their output to foreign markets and of transporting the materials they need from abroad. The cost penalties resulting from geography and poor infrastructure are far greater in Uganda than from high tariffs and other import restrictions.

Southern firms can still break into export markets, however, developing—country firms do export to markets with exacting standards for product quality, reliability of delivery, and consumer safety. Two crucial aspects, however, are often overlooked:

- Non-manufacturing sectors, such as tourism and horticulture, generate significant employment and offer opportunities for supplying increasingly sophisticated products. Although manufacturing is considered more attractive, certain areas of tourism and horticulture can be equally appealing.
- New export opportunities are created as southern producers establish closer links with foreign customers. Producers of labour-intensive products such as garments, horticulture and footwear frequently depend on large retailers and specialist international traders for designs, information about demand and technical support.

Supermarkets make key decisions about which fruits and vegetables to grow, how they should be produced and processed and which firms should be included in the business. Strategic decisions by international producers and retailers in the footwear industry have been crucial in developing new production locations such as Vietnam and Romania. Similarly, work on automotive components production in South Africa and India illustrates how global sourcing by the leading motor companies closes off some markets and opens up others. Export prospects can only be evaluated in the light of global restructuring in these industries.

Emphasising global linkages does not mean that developing countries are powerless in the face of global forces. Even in tightly-structured industries, there is scope for national policy and national strategy. Further more, there are important export sectors that are not structured in this way. Some tourism is dominated by large northern firms and is heavily import-dependent, but there is also enormous potential and national policy will be crucial in shaping the industry and its contribution to the economy as a whole.

For southern firms to break into export markets, certain issues must be addressed, especially in Africa. Some are recognised as important policy issues—investing in human capital and improving infrastructure for example. As one set of constraints are reduced—such as removing policy—induced distortions through trade liberalisation—another set takes precedence. In response to the integration of global markets, southern producers must join the global distribution chains to ensure markets for their exports.

These findings impose hard choices on developing countries. Should a firm allocate limited funds for investment in human capital or investment infrastructure? Future research might contribute by quantifying relative rates of return. On another level, countries may worry about the independence and autonomy of local producers if they are to join a global chain typically donated by northern companies. Rules regulate governmental trade and investment policies but who controls the global buyers and multinational companies whose decisions have such huge impacts on developing countries?

28

# What was Wrong with Structural Adjustment

## *In Defence of a Much-Maligned Strategy*

After decades of stranded development theories, ideologies and paradigms, "structural adjustment", with its demands for clean fiscal policy and an end to uneconomic state enterprises, political privileges, market and exchange rate intervention and corruption, entered the aid arena like a refreshing dawn after a long night of frustrating dreams. Only the "old guard" of planned economy advocates and jealous academicians who had missed the boat were able to shut their eyes to the moral and economic justification of this liberating breakthrough international development policy spearheaded by the Breton Woods institutions then steered by some exceptionally courageous economists.

### Reaction to SAPs

As with any revolution, defeat is awaiting the pioneers at the hands of political power greed, reactionary tactics by the formerly privileged and academic envy. The principal device serving the reactionary forces as a lever of influence on the mood of the "development community" has been the identification and dramatisation of new pockets or strata of (principally urban) poverty allegedly created by structural adjustment measures, while shunning the much broader-based rise in economic activity, real incomes and sense of fair reward in the overall society, especially the rural population. That the hardship experienced by urban poor, formerly privileged under consumer price control and import subsidies to the debit of depressed farm prices or maintained by grossly over expanded public payrolls, was only laying open the camouflaged erosion of the economy and near-bankruptcy of governments and public enterprises, was conveniently downplayed.

These reactionary howls were to be expected. Not that they met the entirely innocent. There had been naively sweeping, overly assuming demands by some structural adjustment missions. But an intellectually vigorous and dynamic "development community" would have coped with the ensuing opposition, strengthened the analytical and monitoring capacities and the political will to endure also rocky roads and bitter medicines on the way to a healthier base. Instead institutional rivalry, political opportunism and emotive populism were thriving. In a way, the "development community" behaved as if it did not want its patient to become able to stand on his own feet and eventually steal its raison d'être.

Worst, the Bretton Woods Institutions themselves, partly under the pressure of the emotive opposition described above fell to the temptation to rescue their lending volume, which was threatened by the frugality dictated to Third World public budgets under structural adjustment recipes, through hardship-easing loans. They thereby corrupted their creation in using it to reinforce their indispensability. As a consequence it soon turned out that some of the most obedient loan takers under structural adjustment terms experienced sharply rising indebtedness, exploited as a disqualifying symptom by the anti-structural adjustment camp.

Whatever the opinions on structural adjustment policies, the commitment to the principles of "good governance" has come to stay, at least on paper, as an almost standard conditionally for official development aid from OECD donor countries. The realisation, matured in the implementation of structural adjustment programmes, that not the quantity of aid, but the quality of Third World government determines the positive or negative course of development, may be regarded as the most valuable fruit of the decades-old policy debate in the 'development community". And the use of aid a pressure or bribing factor towards "good governance" as foreign aid's least disputable purpose.

**Out of the Limelight**

Nothing, however, must be taken for granted. Achievement breeds its challenge! Structural adjustment, though in essence hardly disputable has been pushed out of the limelight and replaced by the oldest actor in the company: eradication of poverty, twinned with an equally perpetual endeavour at the macro-level: debt-forgiveness. This falling back to square one in donors approach to the problems of the south, i.e., the call to alleviate poverty and priorities direct efforts to this end above all other developmental efforts—does it indicate a sell-out of constructive ideas in the "development community"? Has any noteworthy progress been achieved in the past by this approach?

By telling a frugally toiling but independent subsistence farmer that internationally his condition is classed as "poverty", deserving compassion and support by the world community and cancellation of his debts, one can hardly expect a sustainable improvement in his output, satisfaction, or self-respect and even less, when he realises that the help principally provides jobs, fringe benefits and self-importance to a gamut of intermediaries, at home and abroad.

What do those poverty advocates (the"Lords of poverty") really know about the resources, life managements, value systems and ambitions of those they generalize by the billions? The great variance in the conception of life situations, from different external viewpoints.

What the aid system can do for these rural populations classed as "poor"/"underprivileged"/"exploited", is press for justice, i.e., "good governance". The achievement of structural adjustment policy through e.g. abolishing official price and exchange rate distortions, import subsidies and exploitative state agencies, has brought massive income improvement for peasant populations, i.e. the majority of LDC inhabitants, in dimensions unreachable by whatsoever direct "attack" on rural "poverty". What people want is not being benevolently treated as poor, but being justly rewarded for their work, i.e., by access to the unmanipulated market value of their output. Slackening on structural adjustment/ "good governance" conditionally under the present "10 year itch" for paradigm change means foregoing much of the potential opportunities for undoing injustice and exploitation of the masses. It should be clear where priority focus should be placed in ODA policy.

**Small is Not Beautiful**

The direct attack on "poverty", orchestrated by the Bretton Woods institutions under their freshly launched Poverty Reduction Strategy Paper (PRSP) campaign, is being rightly regarded as primarily an NGO domain, since most activities are expected to be carried out at local community level. This would require careful screening and coordinating of NGO activities and their integration via gradual expansion of their experience. But "small" is not "beautiful" for the development financing institution. Disbursement needs are pressing, calling for the new paradigm to quickly provide channels for another wave of loans to the "IDA Countries". Their problem of heavy indebtedness, which would principally exclude most of them from any new loan consideration, shall be solved with one stroke (which only the well-cushioned development bureaucracy can afford); debt relief against presentation of country PRSPs by the respective governments. NGOs are expected of play in the system especially the knowledge gap about the "poor" people's real wants and needs NGOs will naturally be tempted by such expansionary boost to their involvement (referred to sarcastically as their philanthropic empire" by an African conference participant), but this will not be conducive to quality and accountability of their performance, which ideally should be based on private sponsorship in combination with strong target-group provided self-help components.

**Patience and Self-Restraint**

Local knowledge and initiatives cannot be obtained under time pressure. "The grass does not grow faster by being pulled". When will the "development community" learn patience and self-restraint in the approach to LDC's capacity for constructive absorption of aid programmes accompanied by a genuine sense of ownership?

After all these deliberations, how shall development policy be shaped in order to better correspond with reality, without sinking deeper into hypocrisy and frustration?

To come back to the opening question: what was wrong with "structural adjustment"? Nothing was wrong with its intent. In fact this was very right and long overdue. Its implementation, however, lacked patience, perseverance and solid support from the development community, apart from its being corrupted as a vehicle for expansionary lending policy. If aid is meant to not be an end in itself, then structural adjustment policy needs constant reinforcement, underpinned by strict lending discipline. There should be an end to irresponsible lending and easy escape from its consequences by wholesome periodic debt relief burdened on the international tax-paying community. No ODA, either loans or grants, should be made available to governments who are not in active process of implementing "good governance" principles. A monitoring unit, reporting to the donor community on government performance in regard to its "-good government"? Structural adjustment commitment, should be maintained in each and receiving country by "donor consortia" comprising all locally represented bilateral and multilateral development organisations currently extending technical, financial or material assistance to the country.

In order to accommodate the poverty focus without diluting the necessary structural adjustment orientation of ODA, a division of activity-focus between the latter and the NGO sector would seem to be advantageous.

- ODA, limited to the countries abiding to structural adjustment/"good governance" conditionally, with focus concentration on sustainable physical, social and economic infrastructure principally at national and regional level, public management training, higher education and research, consultant and senior adviser services.

- The NGO sector, principally funded by private sponsorship, united to structural adjustment conditionally (but preferably grafted on local self-help initiative), with focus-concentration on the "Third World "poor", i.e. mostly at rural community and low-income township level, for amelioration of living conditions and local resource utilisation.
- Strengthening of linkages between the NGO sector and the UN Technical Agencies to mutual benefit: NGOs in need of professional information, evaluation and advice, of forum for discussion to find an actively supportive window at the agencies; the latter to maintain and develop field contact of research and policy generation, not least as a substitute for their declining project work (giving way to greater concentration on their global functions i.e., serving as information, policy initiation, and coordination/negotiation centre on topics of global concern such as e.g. : human rights, global monetary and trade systems, tropical forest and global marine resources, global and regional health threats, international standards).

In conclusion, it may be called to mind that aid and its institutions have no claim for permanence. They are justified only as temporary functions in a phasing-out process of self-help support. Any claim for unlimited continuity would breed lasting infantilisation.

# 29

# Food Production

During the last 25 years, world agriculture successfully expanded food production faster than population growth. This can continue for the next 25 years and beyond, if appropriate action is taken. Although world food stocks are currently low and grain prices high, the world is not about to run out of food. We can produce enough food for future generation if we choose to do so.

The widespread food insecurity, unhealthy living conditions, and abject and absolute poverty in many developing countries are already threatening global stability. Failure to assure sustainable food security will foster the very conditions that will further destabilise and polarise the world in the years to come with tremendous consequences for all people.

**The Basic Facts**

Poverty is widespread in developing countries, with over 1.1 billion people living on a dollar a day or less per person. Human resource development in developing countries is lagging: 1 billion people lack access to health services, 1.3 billion do not have access to adequate sanitation systems, and one-third of primary school enrolls drop out by Grade 4. Natural resources, upon which future food production depends, are being degraded at alarming rates: almost 2 billion hectares of land have been degraded in the past 50 years: about 180 million hectares of forests have been converted to other uses during the 1980s, marine fisheries are collapsing around the world, and regional and seasonal water shortage afflict many developing countries. Improved appropriate technology is essential to increase productivity. Yet low-income food deficit developing countries are grossly under investing in agricultural research and many are reducing their support.

It calls for sustained action in six priority areas. First, we must selectively strengthen the capacity of developing country governments to perform appropriate functions such as establishing or clarifying property rights, promoting private-sector competition in agricultural markets, and maintaining appropriate macro economic environments. Predictability, transparency and continuity in policy making and enforcement must be pursued.

**Investing in People**

Second, we must invest more in poor people in order to enhance their productivity, health, and nutrition. It is not only unethical but economically wasteful that a large share of the world's population is malnourished, illiterate, sick, and without access to productive resources. Access to primary education, primary health care, reproductive care and family planning information, and clean water and sanitation must be assured for all people. Access by the poor to productive resources and remunerative employment must be improved. Empowerment of women must be supported.

Third, we must accelerate agricultural productivity. Agriculture is the lifeblood of the economy in low-income developing countries. In those countries, it provides up to three-quarters of all employment and half of all incomes. There are very strong links between agricultural productivity increases and broad-based economic growth in the rest of the economy. Agriculture is an engine of growth in low-income developing countries. National and international agricultural research systems must be mobilised to develop improved technologies focused on developing countries, and extension systems must be strengthened to disseminate the improved technologies and techniques. Low-income countries currently spend less than 0.5 per cent of the value of agricultural production on agricultural research compared to 2 per cent spent on agricultural research in middle and high-income countries. An increase of agricultural research expenditures in low-income countries to at least 1 per cent of the value of a agricultural output is urgently needed, with a longer term target of 2 per cent. National agricultural research must be supported by a vibrant international agricultural research system that undertakes research with large international benefit applicable across boundaries. Current investments in international agricultural research are grossly inadequate to provide the support needed by developing countries. It is of critical importance that agricultural research result in reduced unit costs of production. Such cost reductions will make food economically accessible to low-income consumers, and permit producer incomes to increase. To assure relevance of research and appropriate distribution of responsibilities, interactions between public sector agricultural research systems, farmers, private enterprises, and NGOs must be strengthened.

Fourth, we must assure sustainability in agricultural production and sound management of natural resources. Farmers, local communities, and governments must be encouraged to establish and enforce systems of rights to use and manage natural resources, to improve the way water is allocated and used, to reverse land degradation where it has occurred, to reduce the use of chemical pesticides and promote integrated pest management programmes, and to implement integrated soil fertility programmes in areas with low soil fertility. Local control over natural resources must be strengthened and local capacity for organisation and management improved. Investments in less-favoured geographical areas, that is, areas with agricultural potential, irregular rainfall patterns, and fragile soils must be expanded. Most poor people in developing countries reside in rural areas, and most rural poor reside in less-favoured areas. Yet, most investments, including agricultural research investments, still focus on the more-favoured areas. If we are serious about reducing poverty and protecting the natural resource base, the balance between the less-favoured and more-favoured areas must be redressed.

Fifth, we must reduce food-marketing costs in low-income developing countries. The cost of bringing food from the producer to the consumer is very high in many of these countries. Efficient, effective, and low-cost agricultural markets must be developed in order to bring these costs down. Inefficient state-run firms in agricultural in-put markets must be phased out; investment in developing and maintaining infrastructure, especially in rural areas, must be forthcoming; policies and institutions that favour large-scale, capital-intensive market agents over small-scale, labour-intensive ones must be removed; development of small-scale credit and savings institutions must be facilitated, and technical assistance to create or strengthen small-scale, labour-intensive competitive rural enterprises must be provided.

Sixth, we must expand and realign international development assistance. Many years ago, industrialised countries had agreed to allocate at least 0.7 per cent of the gross national product (GNP) to international assistance. Most countries have not reached or do not maintain this target. Not only must the industrialised countries increase international development assistance to reach the 0.7 per cent target, but they must realign it to low-income developing countries. Also contrary to the middle-and higher-income developing countries, the poorest countries are not able to gain access to capital from the rapidly expanding international commercial capital market. Developing countries in turn must seek measures to diversify sources of external funding, stem capital flight; and improve the effectiveness of the aid they receive.

30

## Less Food Security in the South

Combating hunger and poverty is the central point of Bread for the world' s mandate. In our view, that is not so much about the quantity of food produced in the world. On the one hand, it's about its fair distribution and, on the other, the access of poor people to chances of jobs. Put another way, it's to do with access to purchasing power. In the case of agriculture, that is bound up with the question of how food is produced. Whether the technologies applied maximize employment or replace work with capital.

**"Hunger Though Surplus"**

The question of production, employment and distribution are tied closely to the general conditions for development. It is certainly not exclusively external economic conditions, which account for hunger and under-development. Structural deficits, political conditions and wrong policies in Third World countries have become increasingly clear. However, it can still be noted that global economic framework conditions remain enormously important for the development of agriculture in the Third World.

Twenty years ago, Bread for the world publicly expounded the thesis "Hunger through surplus" and had to take much criticism for it—above all from agro-economists. But since then the contradiction between the ever-growing mountains of agricultural surpluses in the northern hemisphere and the increasing dependence on food imports of the South has become ever more apparent. Out of 120 poor developing countries, 107 today are net importers of food.

The North's surpluses of dairy products, grain, beef and sugar—which because of their production costs are exorbitantly expensive—thrust their way on to world market and destroy local supply systems (which are cheap because of subsidies), regional trade flows, and the sales possibilities of potential Third World agro-exporters. Thus, the surpluses contribute to the situation that in many developing countries a policy of neglecting local agriculture can be continued with impurity.

Initially, the promise to work on the yawning gap between hunger and surplus in the world was upfront on WTO agenda. But the pattern of explanation was well simplified. It said that surpluses arose only in those countries, which supported their agriculture positively and, in fact, partly excessively. And that agricultural deficiencies in countries of the South were caused mainly by deprivation of resources and capital. However, the concept of not only reducing neglect of agriculture in the South but also its oversubsidising in the North to a sensible degree and thereby eliminating their distortions of world markets had a great intellectual attraction. At any rate, it promised more justice in agriculture.

**Subsidies Can Make Sense**

To avoid misunderstandings, we have nothing against the support of agriculture in Europe. Above all not when it is done for social, ecological or agriculturally beneficial reasons.

On the contrary, agriculture's important role for food security, the sustainable handling of natural resources, the settlement of rural areas, and the social function of family farms justify a special position for it is economic life, including protection and support.

But that must not be carried so far that surpluses are produced with the help of dubious production methods and then dumped on the world market at markedly less than cost price, causing incalculable damage in the poor countries. On the other hand, purposeful promotion of rural development is a prerequisite and model for greater self-sufficiency worldwide, especially in Third World countries.

### Complementary Functions of World Markets

The poor countries of the South have no alternative than to become self-sufficient in food. The World markets can at best assume complementary functions. The countries would take indeterminable risks if they integrated themselves completely in the world markets, and thereby wanted to make themselves dependent upon global agro-markets. These are and will remain extremely unreliable factors that are conditioned by enormous fluctuations in prices and quantities, the powerful, and in many cases obscure, influences of multinational concerns, the manifold political interventions in the agricultural scene in most countries, and the dangers of social and ecological dumping.

But when we now look at the results of the WTO, we are disappointed. The development question and the balancing of hunger and surplus are finally no longer on the agenda. Programmes to increase food production in the poor countries were not the priority of the negotiations. The liberalisation of agro-policies in the developing countries would have meant making the disadvantaging of their farmers the subject of international negotiations. That did not happen.

On the contrary, the concepts developed with an eye on the reform of agricultural policy in the North, which target the reduction of the support level, are to be transferred to the South without questions. To be sure, there are a whole number of exemptions for the poorest developing countries. But the WTO results have also set the trend there, namely the dismantlement of subsides. We cannot understand how such a thing can be demanded as a policy programme, especially for Africa. Support for African agriculture is largely absent, i.e. there is absolutely nothing to dismantle. That's why many international conferences repeatedly emphasise the need for the countries to achieve a greater degree of self-sufficiency in food by stronger support of their agriculture.

**Agro-Dumping**

Certainly, some changes have been made in the North's agro policy system, which will also have positive impacts on world agricultural markets. However, also here we must express our disappointment. Agricultural dumping will continue. The only difference will be the new policy instrument of direct transfer of income instead of subsidised grain prices. The opening of markets in future will hardly go beyond the current preference conditions.

The entire set of W.T.O. agreements, however, bears the imprint of the two agricultural superpowers, the USA and the European Union, which make mutual concessions and coordinate their agricultural policies. But one hears nothing about the target of freeing the world agricultural market from unnecessary distortions and ensuring justice. The intention of the agro-superpowers was solely to defend their global market shares.

The development aid agencies cannot close their eyes to these problems. On the contrary, in future they must make very much greater effort in suggesting better goals, programmes and instruments which are capable of forming a policy that can then be included in the agenda of the next rounds of negotiations. We may perhaps have slept a bit through the past WTO talks. Therefore it is even more important that we get very much more involved from now on.

31

## Genetic Diversity and Food Security

Maintaining a diversity of crops and varieties is a key to survival for millions of farmers living on impoverished land. For thousands of years, farmers have used the genetic variation in wild and cultivated plants to develop their crops and raise new breeds of live stock. Genetic diversity gives species the ability to adapt to changing environments, including new pests and diseases and new climatic conditions. Plant genetic resources—that component of genetic diversity of actual or potential use to humanity—provide the raw material for breeding new varieties of crops. These, in turn, provide a basis for more productive and resilient production systems that are better able to cope with such stresses as drought or overgrazing and can reduce the potential for soil erosion. The use of genetic diversity—on-farm, through field experimentation or in sophisticated gene transfer procedures—remains arguably the best route so securing our food and that of our children.

Although science has made enormous strides in improving the world's ability to feed itself over the past three decades, we cannot afford to rest idle. Nearly 800 million people in the developing world do not have enough to eat. In these regions, the rural poor represent about 73 per cent of the people living in poverty. They often live in marginal or unsuitable farming areas, such as zones with saline soils, and conditions, or degraded or hilly areas. Often isolated from other farms and far from urban areas, many poor farmers have barely benefited from agricultural developments elsewhere. In many cases they do no have access to commercially bred high yielding crop varieties. Diversity flourishes and remains important under such conditions.

**Selections and Breeding**

Poor farmers are well aware of the relationship between the stability and sustainability of crops and crop varieties on their lands. Their management and use of a diverse range of plants has often helped them to survive under the most difficult conditions. By growing a range of different crops, farmers have a better chance of meeting their needs. These might be crops that mature at different times or that can be easily stored to help to ensure a stable food supply throughout the year. They may also help farmers provide a nutritionaliy balance diet for their families, exploit different environment niches that exist on their land, or diversity their income sources.

Importantly, the genetic diversity contained in different varieties provides farmers with options to develop, through selection and breeding, new and more productive crops that are resistant to pests and diseases. The result may be a vast range of local varieties of crops grown by farmers in any one area.

Not respecting diversity can incur high costs: in 18th century Ireland, where potatoes were the only significant source of food for about one third of the population, farmers came to rely almost entirely on one very fertile and productive variety, which proved susceptible to the devastating potato blight fungus. The resulting famine caused the death or emigration of more than 20 per cent of the population.

The value of diversity goes well beyond its ability to support stable production systems in marginal environments. As the world's human population rises, environmental problems (desertification, deforestation, erosion etc.) are intensifying, Climate change, particularly global warming, could bring about drastic changes in the location of the world's agro-ecological zones. Farmers will require new crop varieties capable of producing under diverse conditions, without adding ever-increasing amounts of fertilizers and other agro-chemicals. Because of the limited scope for growth in the world's cultivated areas, each new generation of varieties will have to be more productive than its predecessors.

Much has been written about the use of genetic engineering in plant breeding. Modern molecular techniques can be used to transfer genes from one living organism to another or to change the genetic material within to produce more desirable traits. Genetic Engineering has enormous potential to help solve problems that have proved intractable using conventional breeding approaches, such as developing crop varieties with in-built resistance to keep pests and diseases and tolerance to stresses such as drought. However, the possible impact of these techniques, particularly on human health and the environment, is giving rise to fierce worldwide debate.

Take the case of banana and its close relative plantain, two of the developing world's most important crops. Their improvement is hindered by the sterility of most cultivars, a problem that can be addressed through genetic engineering. It is now possible to transfer gene constructs, such as those associated with disease resistance, directly into varieties with other desirable characteristics, drastically reducing the need for pesticides.

Today, research on genetic engineering is focussed on the development of commercial varieties of the world's major crops of interest to industrialised farmers. Many of the staple crops of importance to poor farmers in developing countries, such as cassava, bananas, beans and yams, have received relatively little attention. This situation is likely to continue as plant breeding is increasingly privatised and biotechnology becomes the fast-growing province of private industry. Meanwhile, the high costs of the new technologies are quickly exceeding the capacity of many, if not most, public research institutions—both in developing and developed countries—to support them. Thus, for the time being, increasing agriculture's role in the development of the world's poor is likely to continue to depend on the identification, maintenance and use of genetic diversity.

## Employment and Poverty Alleviation

Today the key socio-economic problem is large-scale unemployment. Spreading joblessness brings many other problems in its wake. It erodes national income and living standards, aggravating the already grindingly difficult job of promoting development and alleviating poverty. Joblessness also raises government budget deficits, increasing macro-economic instability while soaking up investment for productive capital expenditure, education, training and relief aid. And joblessness ruins lives and communities by depriving people of the dignity and satisfaction that comes with earning one's keep and making a contribution to the well being of family and society.

Theories about how best to nurture development (and thus create jobs) have shifted considerably over the last decade. The state role has evolved, in the minds of many, from being a source of relief for the problems of unemployment, poverty and underdevelopment, to being a fundamental cause of these problems through the distorting impact of its intervention on the market.

However, the more market oriented philosophy that grew up during the 1990s has yet to provide convincing solutions in practice at least not on a grand scale and especially not in terms of job creation as the present jobless economic recovery demonstrates.

The weakness of the current recovery and past approaches to economic development can be traced to the failure to consider employment as the predominant means of promoting growth and alleviating poverty. In policy circles it has too long been an almost ignored priority.

Current trends thus bode poorly, particularly as unemployment rates soar. In light of the circumstances, we need to begin re-examining some of the fundamental questions—if only to find out what has gone wrong with the answers.

**Minimum Wage?**

Let's begin with wages. With corporate restructuring in full force on a global scale, are low wage rates required to raise employment and maximize profits? A top manager of a multinational consumer electronics group certainly thinks so; he likened the perfect factory to a ship "so that we could move it around the world to where labour was cheapest". Perhaps, but this bottom-line emphasis on unit labour costs ignores at least two other factors; namely, that higher wages can act as a screen to select more productive workers and that higher wages translate into better productivity via improved worker nutrition, increased consumption and a generally healthier quality of life.

If higher wages bring these benefits (and it is an open question) should government insist that there be a minimum wage rate? Neo-classical economists tend to respond "no", assuming that a higher wage rate puts money into the pockets of some low wage workers while forcing many others out of work because companies cannot afford to pay them.

**Technology Transfer**

The impact of technology is another area in need of study. Technological innovation is usually labour-saving and tends to originate in industrialised countries, moving toward developing countries like India, Pakistan where labour tends to be low cost and abundant. Would it therefore make sense to slow down or somehow restrict technology transfer, especially to development markets, in the interest of preserving employment?

The answer here is clearly—no. Historical evidence abundantly demonstrates that attempts to retard technological progress bring about grater poverty and lower growth. Technology, infact, is at the heart of the new endogenous growth theory which is very much in vogue among development economists today. Slowing down or inhibiting technology transfer would certainly dash many countries' development hopes and aggravate poverty. However, the relationship between technology, development, employment and poverty alleviation is not without its complications.

In the 1980s, the buzz word among development specialists was "appropriate technology", i.e., small-scale and labour-intensive technologies that would increase productive output while allowing an equilibrium solution to be found such that the ratio of the productivity of labour to that of capital is proportional to their relative prices. The conditions for this "small is beautiful" approach to technology tended to be best met in agricultural production. However, where manufacturing industry is concerned, the small-is-beautiful approach foundered badly when the only viable technological alternatives proved to be highly capital-intensive.

**Development Gap**

A wide gap has emerged between developing countries with an inward focus (which tended to be projectionist and pursue policies of import substitution) and those with an outward focus and a policy of pursuing export-led growth. Competing in international markets requires technology that is as good as or better than that found in advanced, industrialised nations. Small, therefore, is not beautiful in the global manufacturing economy where product standards are high and the elasticity of substitution between labour and capital is very limited.

The drive to obtain state-of-the-art technology thus leads to a policy conundrum: it is a pre-condition for success in manufactured exports, but the impulse to compete successfully in this most lucrative sector speeds up the transfer of technology from the developed to the developing world, thus reinforcing the bias toward labour saving equipment in developing countries and accelerating a process that is seen as a source of job loss in the industrialised countries.

**Technology and Jobs**

Before concluding that modern technology transfer is inimical to employment in developing countries, we have to distinguish clearly between technology's static and dynamic consequences. In a static sense, it is true that highly capital-intensive export industries may not create much employment on a net basis, but the dynamic effects of technology transfer do contribute to economic growth. And growth, in turn, generates multiplier effects in the form of demand, which stimulates ancillary production activities (like food processing or consumer goods) that rely on more labour-intensive technologies.

The problem is that the diffusion and application of technology on a global scale blurs the categories of international product specialisation and creates a much more competitive and conflict-prone international environment.

For example, we have already seen the Asian Tigers move from producing goods such as textiles and processed food to producing hi-tech and value-added consumer durables. This advance is only possible due to the growth of human capital (facilitated by investment and higher incomes) and it leaves production of textiles to other industrializing countries, like Indonesia, the Philippines and now China. But the dynamic comes at the expense of jobs in industrialised regions, like the US and the EC, which lost more than a quarter of their work force in textiles during the 1980s. In spite of job losses, advanced countries continue to produce textiles, notwithstanding major differences in the hourly wage rates for spinning and weaving and the fact that essentially the same hi tech equipment is being used in most production centres.

**Protectionism**

What has happened in textiles is happening in other industrial sectors (automobiles, for example) as well. The intense market competition is providing to be a source of trade conflicts, and possibly protectionism, as jobs come under increasing pressure.

For many workers and managers, the benefits of foreign direct investment look increasingly like a zero-sum game for employment, and there is a real risk that the tenuous link between overall growth and employment will break down altogether. It is hardly surprising that we are already seeing negatively affected workers and local businesses clamouring for protection in advanced countries.

**Governments Role**

The concerned governments are suppose to carry out much of this research. The three initial lines of inquiry follow from three reasonable assumptions about the future.

- First, increase in welfare and consumption subsidies are out; investments in training and human capital are in. How can investments in human capital be directed to positive employment effects? Is it perhaps not time to explore more fully benefit schemes targeting the unemployed and the unskilled poor providing them with the type of subsidies that would enhance their human capital, improve their health and productivity through better nutrition and preventive medicine, and restore the dignity of holding a job?
- Second, given the quasi-inevitability of increased automation in manufacturing, how can other sectors (particularly agriculture and services) be developed to export their long-term potential for employment creation?
- Third, given the inevitable pressures of work and productivity in the global economy, what sort. of alternative institutional arrangements need to evolve with respect to industrial relations, employment and work conditions?

Finding answers to these and other questions will require no small amount of new thinking, but parochialism or a failure of imagination would be fatal flaws in this global era.

33

# Aid Effectiveness as a Multi-level Process

Parallel to the widespread decrease of aid resources provided by donor countries to developing countries in recent years, debate and research on how to make aid more effective has become a major concern. Usually, it is suggested that decades of development assistance have at best produced marginal results in terms of improving development levels in the South. Little mention is made of donor's policy shortcomings and the negative impact of these on efforts aimed at reforming and redefining development cooperation in order to enhance aid effectiveness. The policy parameters and operating frameworks of existing and policies continue to inhibit higher degrees of aid effectiveness. In many donor countries, opinion polls indicate waning public support for development aid.

Increasingly, the moral case for aid is called into question and deeper world market integration tends to be seen as the panacea to continued economic decline and social destabilisation in the South. Against this background, cooperation between donor and recipient actors is faced with a duel uphill struggle. First, fewer resources can be mobilised to meet growing developmental needs. On the other hand, to organise and manage development policies and programmes in a results oriented manner, grows more difficult. The threat of further aid cuts and of further drops of public support for providing aid become ever more real. A closer look at the organisational complexities and political constraints under which development cooperation is expected to perform effectively may help to improve current aid management approaches.

**Towards Conceptual Clarity**

At first sight, catchy definitions of what constitutes effective aid might appear attractive to use, in particular with regard to economic indicators, The term "aid effectiveness" is easily used in the same vein as "efficiency", "significance" or "impact" of aid. At times, obsession to measure and demonstrate the results of aid supported development processes can be observed among policy-makers and administrators on the donor side. Still the understanding of aid and its effectiveness as being part and parcel of a cooperation relationship between donor and recipient side parties, is scarcely embedded in practice. To determine how to make aid more effective requires more than a quick impact analysis of an individual and perhaps even isolated development project. Consequently, defining the concept of aid effectiveness needs to take into account at what levels cooperation is focussed on. To strive for sustainable and effective modes of development cooperation will entail the need to combine recipient ownership of the development process with donor accountability concerns.

Performance expectations cannot be exclusively placed on the recipient while donor interests, their aid management systems and procedures remain unchanged.

An extended and more analytical, process oriented definition should take into account four main aspects of aid effectiveness:

*(a)* Effective aid must relate to the building and/or strengthening of in-country aid management capacity:

*(b)* To maximise the degree of aid effectiveness, local ownership of the aid process is essential: from setting of priorities through policy formulation and implementation on to the evaluation stages of the process;

*(c)* Increasing recipient side capabilities to take charge of aid relationship, will need to be combined with arrangements to meet legitimate donor accountability concerns;

(d) Aid effectiveness is a two-faceted objective: its realisation is equally dependent on increased transparency of donor motives and on dropping of non-developmental, political and economic aid objectives of donors.

In addition a broader range of stakeholders in the aid relationship needs to be actively involved: extending beyond accountable government and implementing agencies, to include democratic institutions and organisations of civil society and of the private sector.

Applying any definition of aid effectiveness without disaggregating macro-economic data and taking into account country specificity will only lead to unhelpful generalisations about aid and its effectiveness. It would seem more appropriate to adopt working definitions against which to assess effectiveness of aid resources at a country-specific level. On such a basis one could expect to arrive at more reliable indicators of how well aid resources contribute to improving developmental standards and meeting existing needs.

**From Definition to Success—Key Requirements**

Having reached agreement between the recipient and donor on what should constitute effectiveness of aid is only a starting point. Embarking on democratic, peaceful and participatory patterns of economic and social development must follow: to arrive at significant and lasting improvement in many of the least developed countries will be a long-term process. This being said, it is crucial to design and implements such forms of development cooperation which involve a wide range of recipient side actors, not only from the government side but also from civil society at large. Seen as a process of increasing inclusion of intended beneficiaries of aid, the commitment to decentralise as well as entrust aid and its management grows in importance.

To fully capture Third World development realities, policy frameworks inspired by neoliberalist-type of development concepts and theories are grossly inadequate. The views and positions on aid articulated in the World Bank and the IMF, or in many if not most bilateral aid administrations in OECD countries, represent only one side of today's international cooperation, namely the donor side. The major weakness to point out with respect to this locus of debate, is a profound under representation if not even a total absence of recipient experiences and perceptions on aid in general and on its effectiveness in particular. There should be little doubt that ignoring to not actively identifying and involving such perceptions, leads to strongly donor driven aid.

To circumvent recipient side insights and views on strengths and weaknesses of aid strategies and mechanisms, will result in limited local commitment and sense of ownership over the aid process. Mutual decision-making between donors and recipients remains a rare policy approach. Aid procedures that are based on local management and less control-oriented donor roles in the aid process are still exceptions in development cooperation.

Structurally, in terms of the policy environment within which development aid is expected to function, the overriding policy framework in general based on structural adjustment policies (SAP). But the underlying conclusion made by proponents of SAPs that these policies induce aid effectiveness, has yet to be proven valid. It must suffice at this point to emphasize that there is not a priori relationship between world market integration under structural adjustment and sustainable development in poor countries. Aid to these countries which is solely intended to reinforce fundamentally uneven and unequal patterns of world market integration should at be scrutinised critically.

Some central issues need to be addressed in the course of improving aid and its effectiveness:

- institutional dimensions of aid relationships require strong policy-attention, both on the donor and the recipient side;

- capacities to effectively identify and formulate aid priorities need to be strengthened in recipient countries;
- local capacities to sustain reform efforts must be reinforced.

**Levels of Intervention**

If the design of aid and the terms upon which it is provided to a developing country are largely determined by the donor, the aid relationship can be characterised as essentially hierarchical. Recipient side views will rarely surface, as they are either not identified, or not well formulated. Possibilities of a recipient-led development strategies can be limited. Unless scope is provided to the recipient side actors to assume responsibilities, aid effectiveness is likely to remain low or fluctuating, and the sustainability of donor aid efforts will remain doubtful.

National planning processes and courses of national development in recipient countries should be seen as most effective where they are led under local responsibility and control. To arrive at this ideal situation, gaps need to be reduced and closed at the various intervention levels.

Donor aid resources provide valuable support for this process. Their effectiveness in meeting long-term objective of aid will need to be assessed on the basis of how well they perform at the different levels. Individual donors will expectedly perform differently at the various levels. What will prove to be the ultimate test for effectiveness is how well the donor aid performance accomplishes the broader objectives of development cooperation and how well it includes sustainable results.

In the analytical frameworks outlined here, development cooperation would seem to be confronted with the effectiveness gaps at the:

- Structural level: International trade and investment patterns, debt problems and world market integration process appear as long-term constraining factors upon aid and its effectiveness;

- at the policy level, dialogue and partnership in development cooperation are instrumental factors in recluding planning and co-ordination gaps with regard to policy analysis and formulation;
- the institutional level is where pertinent capacity gaps exist: capacity development efforts of donors and technical assistance measures play an important role in addressing weaknesses in aid effectiveness within a country's institutional setting;
- finally, at the level of aid projects (programmes), it is generally the lack of sustainability of aid interventions which causes development activities to falter once donor support decreases or stops. In addition to technical cooperation, financial and material inputs serve to maintain project momentum and goal realisation: the issue of how to develop local capacity sufficiently in order for indigenous organisations to continue project activities initially supported by donor aid, remains the most important issue to address at this level.

**Fostering aid Effectiveness**

Donor and recipient development efforts are too often isolated from one another, or poorly coordianted. They fail to address managerial and implementation bottlenecks. Cross-sectorial linkages, as well as interdisciplinary approaches to aid problems are only slowly gaining ground. It is increasingly obvious, that decisions on aid issues are subjected to concerns outside of the responsible ministry: finance ministers, and unfortunately even defence ministers have a strong say in how much aid is to be provided, where it is to be concentrated and under what terms to be utilised. Inside of recipient countries, large portions of national budgets are allocated to non-development priorities with little or no impact on alleviating urgent poverty problems.

Development cooperation may make the biggest impact and be executed most effectively where donors and recipients agree upon multi-level aid strategies. To give an example: building a road to a remote rural area may well be done in an effective project manner: it is equally important to have a functioning transport authority in place to ensure maintenance of the roads. If this authority operates within a nationally defined infrastructure policy, best in accord with national trade and investment priorities, then the effectiveness of the project-level road building programme has a good chance of being high.

Institutional changes to set the stage for a profound reform process in development cooperation are needed. Reprioritising national budgets to reflect identified in country development needs may be one step. Setting up policy evaluation and formulation units can be complimentary measures. Deregulating markets and investment rules may serve to please donors, but dumping of cheap products which strangle local production efforts may easily result. Regional cooperation, including intensified South-South cooperation can provide some counterbalance. There are only a few areas where changes in the current system of development cooperation can occur, with a view to better manage the complexities of aid and the social, cultural, economic and political backgrounds against which they take place. The will and commitment to take policy action in both donor and recipient countries, through the broadest range of stakeholders and institutions as possible, will be the test for genuine efforts at improving development relations between North and South and organising cooperation effectively.

34

# Employment and Promoting Ecology

## *How a Service Culture Could Put People Back to Work*

We are facing two big and urgent social problems: employment and Ecology. Both the unemployment of millions of people and the progressive destruction of the ecosphere are alarming. But they are linked with each other. The 'greening' of industrial products, processes and services could provide many more jobs.

Unemployment has many causes, including:

- Sluggish markets;
- Stagnating or declining purchasing power;
- Growing uncertainty about the future at all levels;
- lack of will and/or ability to innovate.

But joblessness is by far due mostly to the high efficiency of industrial machinery, which produces ever more, ever faster, with ever fewer workers.

**Waste of Resources**

The extremely high productive use of human labour and the extremely low productive use of resources are manifested by gigantic mountains of waste. Already today, the junked cars on scrap heaps alone would form a line that would reach to the moon. The scene is the same with discarded electrical and electronic appliances. Every year, millions of tons of ovens, washing machines, refrigerators, dishwashers, TV sets, entertainment electronics equipment and small appliances are being wasted.

If we throw away all these things after a relatively short time we are not only being wasteful and irresponsible with resources, but equally so with people's work. For with the products and materials we discard, we also dispose of the human labour they contain. It is imperative that we radically reduce the enormous turnovers of material and energy. In other words, the productivity of raw materials and energy must be markedly increased. Specifically, that means we must draw as many services as possible from one kilogram of material or 1 kWh of energy. Reducing the enormous flows of materials into the industrial system, as well as developing cycles of materials and responsibility (the manufacturer taken back and repairing and/or remanufacturing used products and materials) and the main pillars of a sustainable development that can cope with the future.

The industrialized nations must cut their consumption of raw materials by a factor of about 10 by 2050; if they are to be able to handle the challenges of the future. To achieve that reduction, innovation efforts must be directed at increasing resource productivity and/or ecological efficiency. In particular, strategies to extend the useful life of goods and intensify their use could result in reducing both the speed and volume of the flows of resources to industry.

**Increasing Resource Productivity**

In dealing with nature, we and industry are facing radical change. This is the transition from environmental protection (preservation of nature and health) to greater resource productivity (which at the same time means greater competitiveness). As a rule, environmental protection costs money, while higher resource productivity usually cuts manufacturing costs and/or increases a company's profitability. If the company can sell the same utility or benefits while using fewer resources, it saves twofold: in buying raw materials and on waste disposal. Thereby the rule is that goods and components cycles are more profitable than resources cycles, and that the company which is first in the market gains an additional competitive advantage in terms of a lead in knowledge and image. If a service, or benefits in the form of services, can be sold instead of products, the decoupling of company success and materials flows is even greater.

**Impacts on Employment**

The two social problem areas of work and ecology have to date been perceived and treated separately in politics, in industry and in our own minds. And, I believe, with little result. The link between the two must be established.

The strategies to boost resource productivity would have considerable impacts on the change in industrial structures, on handling existing product inventories, and on employment. In particular, the strategies would lead to a switch of focal point from a raw materials-intensive and use-value-related service economy. This is where another view of profitability comes in. Business management would no longer focus on value added, but on maintenance of value over longer periods based on the intrinsic value of a product. Expressed as a question, the value factor, which would move to the centre of business thinking and dealing, means: how can the utilisation value be improved and sold? How can products be made with as few raw materials and as little energy as possible and create a high benefit as pollutant-free as possible for as long as possible during their entire life-cycle?

With regard to employment, the production of long-life goods would appear at first sight to lead to a reduction in the need for work. In fact, however, the strategies to increase resources productivity have positive net employment impacts. The reason is that saving resources is based in principle on substituting energy by work, rather than the reverse as has been customary to date.

If the useful life of products is extended, that will not only preserve most of the materials and energy they contain as well as the work invested in them. The products will also require a considerable amount of mostly skilled work input. Reconditioning products is as a rule more labour-intensive than manufacturing them. So large-scale reconditioning and repair work increase the number of skilled jobs and at the same time reduces the inflows of materials and energy.

Comparing a car with a life-cycle of 20 years with two others that each have useful lives of 10 years gives a good example. The first car causes an increase in employment per life-year of about 50 per cent in terms of total work input in manufacture, service, repairs and reconditioning while at the same time reducing the energy consumption by half.

**Regionalisation of Industry**

Extending product service life would also mean replacing energy and/or capital by skilled work, helping to save money to boot. But not only rising costs of disposal, materials and energy would reduce consumption. Increasing transport costs would also mean that carrying all kinds of freight halfway around the world would make less and less business sense. That would result in ever more products and materials being circulated, reconditioned, and recycled or reduced on a regional basis. In turn, that would create regional jobs, and be more profitable as well as more productive of technology—not only from ecological aspects.

In addition, a way of doing business which encompassed material and responsibility cycles would no longer differentiate between manufacturing and reconditioning, or between marketing and remarketing. The structure of such an economy would be predominantly decentralised and regionalised so that it could adapt itself to the new cycles. It also would benefit from the greater efficiency of the new working practices.

True, jobs would be lost in the sectors of central production, and raw materials extraction and processing. But at the same time, more and higher-skilled jobs would emerge. These would not only be better qualified jobs, but also decentralized because reconditioning, repairs and maintenance must be done near the customer. And that, in turn, would also reduce goods traffic.

In addition, skilled workers would be needed because in many cases of small production runs it makes sense and is also more economical to hire such people. They can work faster and more flexibly—and mostly cheaper-than fully-automated production lines.

There also would be a growing need for maintenance, repairs and reconditioning. More and more people would be wanted for reconditioning, that is, the remanufacturing of old products. As reconditioning involves far more craft work than highly rationalised new production, there would be a positive impact on the labour market if there were more of the former and correspondingly less of the latter.

**From Production to Services**

Switching to long-life products and changing from selling products to selling use-values would strengthen the current trend of jobs shifting from industrial production to the service sector. For example, if the service of individual transport were to be sold instead of the product car, the company with the competitive advantage would be the one that had a service centre in every town and village, with appropriately staffed workshops and sales or rental facilities.

Enduring change towards a knowledge-intensive and use-value-related service economy would not only mean that more people would be needed to fill jobs. It would offer more opportunities for part-time work, as well as possibilities of employment for older people and the handicapped. People who earlier could not keep up with the pace of working life would be more inclined to return to it. Another impact would be that many companies would reduce their dependence on the world market. They would no longer switch certain tasks abroad, but assign them to their part-time employees, helping them to meet their commitments as self-employed entrepreneurs.

The latter would be accommodated by an ecology-driven fiscal reform which would make massive cuts or changes in subsidies and raise the cost of energy and raw materials consumption. This move would be accompanied by a reduction in income tax and non-wage costs such as social security contributions. The market would thus be more efficient, energy- and material-intensive new production more expensive, labour intensive repair work and reconditioning cheaper, and jobs would remain in the home country or region.

A number of more recent studies show clearly that an ecological tax reform would help to create jobs, and thereby could make a decisive contribution to reducing unemployment.

35

## Population Growth and Climate Change

Over the last half-century, carbon emissions from fossil fuel burning expanded at nearly twice the rate of population, boosting atmospheric concentrations of carbon dioxide, the principal greenhouse gas, by 30 per cent over preindustrial levels. All major scientific bodies acknowledge the likelihood that climate change due to the buildup of greenhouse gases in the atmosphere is indeed under way. The 15 warmest years on record have all occurred since 1979, and 1998.

The destabilisation of our climate threatens more intense heat waves, more severe droughts and floods, more destructive storms, and more extensive forest fires. The related shifts in rainfall and temperature may jeopardize food production, the Earth's biological diversity, and entire ecosystems, as well as human health by expanding the ranges of tropical diseases. Unless efforts to curb them are stepped up, carbon emissions will continue to grow faster than population over the next 50 years, driving the Earth's climate system into unchartered territory. The Intergovernmental Panel on Climate Change (IPCC) estimates that an eventual two-thirds reduction in global emissions is needed to avoid precariously high levels of atmosphere carbon dioxide concentrations.

The IPCC and U.S. Department of Energy (DOE) project that emissions from developing countries will nearly quadruple over the next half-century, while those from industrial nations will increase by 30 per cent. Although annual emissions from industrial countries are currently twice as high as from developing ones, the latter are on target to eclipse the industrial world by 2020.

Higher per capita carbon emissions accounts for roughly 55 per cent of the increase in emissions projected for developing nations. Emissions per person are due to more than double from 0.51 tons of carbon per year in 2000—just one fifth of the industrial level—to 1.14 tons in 2050. The remaining 45 per cent of emissions increases is due to population growth.

Fossil fuel use accounts for roughly three quarters of world carbon emissions. As a result, regional growth in carbon emissions tend to occur where economic activity, and related energy use, is projected to grow most rapidly. Emissions in China are projected to grow over three times faster than population in the next half-century, as emissions per person soar from 0.77 tons of carbon to 2.81 tons due to booming economy that is heavily reliant on coal and other carbon-rich energy sources. In Africa, in contrast, emissions per person are expected to scarcely change-growing from the currant level of 0.33 tons in 2050, despite a threefold increase in total emissions.

The effects of population growth are most profound in countries where people are heavily emitters. For example, the 115 million people added to the population of the United States between 1950 and 1998—an increase of nearly 75 per cent in just 45 years—account for more than one tenth of current global emissions. And the carbon emissions of the 75 million people who will be added to the U.S. population in the next 50 years roughly equal the emissions of the 1.3 billion people who will be added to Africa during that period.

Deforestation and other land use changes account for the remainder of world carbon emissions. Forests have served as a sink for carbon throughout much of human history. In recent years, however, the world's forests have become net sources of atmospheric carbon, largely due to forest burning and clearing in the tropics. Six months of fires in Asia in 1997 and 1998 released more carbon than Western Europe emits from fossil fuel burning in an entire year. The carbon contribution from this source will likely increase in coming years as the burgeoning human population continues to cut down forests.

36

# Biodiversity

As human population has surged this century, the populations of numerous other species have tumbled, many to the point of extinction. Indeed, we live amid the greatest extinction of plant and animal life since the dinosaurs disappeared some 65 million years ago, with species losses at 100 to 1,000 times the natural rate. But humans are not just witnesses to a rare historic event, we are actually its cause. The leading sources of today's species loss, habitat alteration, invasions by exotic species, pollution, and overhunting are all a function of human activities.

Human activities have pushed the percentage of mammals, amphibians, land fish that are in "immediate danger" of extinction into double digits. The principal cause of species extinction is habitat loss—the result of encroachment by humans for settlements, for agriculture, or to claim resources such as timber. A particularly productive but vulnerable habitat is found in coastal areas, home to 60 per cent of the world's population. Coastal wetlands nurture two thirds of all commercially caught fish, for example. And coral reefs have the second highest concentration of biodiversity in the world, after tropical rainforests. But human encroachment and pollution are degrading these areas: roughly half of the world's salt marshes and mangrove swamps have been eliminated or radically altered, and two thirds of the world's coral reefs have been degraded, 10 per cent of them "beyond recognition". As coastal migration continues—coastal dwellers could account for 75 per cent of world population within 30 years—the pressures on these productive habitats will likely increase".

Habitat loss tends to accelerate with an increase in a country's population density. This is bad news for the world's biodiversity hotspots-species-rich ecosystems at greatest risk of destruction. Twenty-four of these hotspots, containing half of the planet's species, have been identified globally. Some of the most important hotspot countries will reach population densities that have been linked with very high rates of habitat loss. Five of the six most biologically rich countries could see more than two thirds of their original habitat destroyed by 2050 if this historical relationship holds.

Related to loss of habitat is the growing incidence of plant, animal, insect, and microbial invasions of ecosystems worldwide as human interchange increases. These "exotic species" sometimes dominate local ecosystems, eliminating native species and reducing overall diversity. Exotics are implicated in 68 per cent of all fish exticntions in the United States this century, for example. Growth in human travel and commerce explains many accidental invasions by exotics, but foreign species are also deliberately introduced into farms, plantation forests, and aquaculture systems. Although only 1 per cent of exotics cause widespread damage, exotic species are the second leading cause, after habitat destruction, of species loss worldwide.

Other, often diffuse effects of expanded human activities also disrupt ecosystems. Nitrogen, for example, is now made available to plants at more than twice the preindustrial rate as a result of fertilizer production, cultivation of nitrogen-fixing crops, and the burning of fossil fuels. This overfertilisation of the Earth favours some species at the expense of others, leading to a reduction in diversity and resiliency of land and aquatic ecosystems.

Likewise, greenhouse gas emissions could disrupt ecosystems on a vast scale. As with nitrogen, increased levels of atmospheric carbon may favour some species over others: annuals over perennials, for example, or deciduous trees over evergreens. To the extent that greenhouse gases induce changes in global climate, many species may be at risk as habitats shift or shrink, and as some life forms, such as insects or animals, adapt and migrate more quickly than others, such as plants. And as sea levels rise with a change in climate, ecosystems such as coastal wetlands could be destroyed.

37

## Democracy and Poverty
*Are they Interlinked?*

Democracy assistance and poverty reduction are rightly becoming two focal—and related—issues for development assistance. Increasingly, many organisations, including intergovernmental, national and civil society, are focusing their work on these two areas. Futhermore, the relationship between these two issues is complex and ever changing. There is thus a need to develop methodologies of linking democracy assistance and poverty reduction at both the policy and programme levels. International IDEA (Institute for Democracy and Electoral Assistance) in cooperation with the World Bank and the United Nations Development Programme, is developing concrete strategies that address these two objectives in a mutually reinforcing way. Through an overall situation analysis followed by regional meetings in sub-Saharan Africa, South Asia, Latin America, the Caucasus and the Arab region, the Institute has marshalled evidence of some of the key problems that affect democracy consolidation and poverty reduction in these countries:

- Corruption and its undermining effect on popular confidence in public institutions;
- Continuing economic instability coupled with the lack of strategies for addressing the twin challenges of poverty and increasing popular participation in its alleviation;
- The extremely limited nature of citizen's influence on overall policy and decision-making processes despite the spread of formal democratic institutions;

- A trend in many post-communist states towards viewing growing poverty as a direct consequence of a transition to democracy.

In short, the evidence is not very encouraging for the prospects for democracy consolidation and poverty reduction. The critical step, International IDEA, advocates is the development of an approach that not only seeks to put democracy assistance and poverty reduction on top of the development assistance agenda, but also to encourage all involved to treat them as twin elements of an integrated programme of action.

Through a focus on accountable governance, promotion and protection of citizenship and rights and increased popular participation, International IDEA believes that both democracy and poverty reduction can be addressed simultaneously. Policy recommendations are being developed and will be shared in the course of this year with governments, international organisations and civil society bodies.

International IDEA believes that democracy promotion can be used as a tool for fulfilling a variety of objectives. Democracy matters because it protects human right and preserves human dignity. But democracy also matters because it helps to address some of the most critical challenges facing states today: peace, development, economic growth and stability.

Democracy does not guarantee any one of these, but increasingly it seems to be a precondition for them in the long term. Thus, advocating democracy goes beyond being a moral issue; it becomes *fundamental* to advancing the well-being of people and the stability of states. International IDEA will continue to explore the link between democracy and the major issues facing society today and continue to argue the case democracy.

38

# Major Cyclones in Andhra Pradesh

## *Some Observations*

Cyclone is considered as the most devastating of all natural phenomena. Repeat cyclones create no sense of security to the wealth and life of the people. It is also well known that the sudden and rapidly developing cyclonic storm not only disrupts the prevailing order of life and produces danger, death and loss of property to large number of people residing within a geographical area but also creates environmental imbalances.

Andhra Pradesh is one of the states facing moderate to severe and repeated cyclones affecting the life of the people of the state and especially coastal districts. In the history of Andhra Pradesh, there is a long list of cyclones. The cyclone which hit the town of Machilipatnam in the year 1864 was one of the severest in the history of the world. The tidal wave was so vast and rushed with such force that it submerged the coastal areas deep up to twenty miles. It took many weeks for the flood water to sink. The entire area, which was submerged under seawater, became saltish and consequently infertile for many years to come. It appears that in the year 1796 also Machilipatnam and surrounding areas were hit with a big tidal wave which claimed over 20,000 people besides loss of property worth more than Rs.250 crores then. The impact of cyclone of 1864 was more on life and property of the people. More than 35,000 people and 3 lakh livestock dead and worth of nearly Rs. 600 crores property was lost. Almost all the houses at Machilipatnam were completely ruined. Even the mighty fort of Britishers turned into shambles but for one building and the Bell tower of the church. With this calamity and Machilipatnam being prone to frequent cyclones, the Britishers shifted all their establishments to Madras, after this cyclone. The Britishers erected two monuments in memory of the 1864 cyclone victims—one of Robertson Square in the heart of the town and the other at Machilipatnam Fort.

Relief measures were taken up to persons who were deprived of the essential needs of life because of natural disaster resulting from cyclone of 1864. Cyclone relief consists largely of emergency provisions of food, clothing, medical care and shelter with the aid of voluntary contribution from other communities and countries. Even International Red Cross and other voluntary organisations did not take cyclone relief as one of the chief activities. Only in 20th century these organisations have been extending assistance to the victims besides greater role of the central and state governments in relieving the population affected by the cyclones.

The second major cyclone in the state like that of 1864 cyclone, hit the coastal region on the night of 19th and 20th November, 1977. There was terrific gale with speed ranging 120-150 Km. per hour. Coastal districts of Andhra Pradesh such as Krishna, Guntur, East Godavari, West Godavari and Prakasam and to a small extent Srikakulam, Nellore and Visakhapatnam districts were affected. The most affected areas were Divi, Machilipatnam, Bapatla, Repalle and to some extent in Chirala Taluk. Huge tidal waves engulfed the coastal region of the Divi, Machilipatam and Repalle Taluks. There was extensive damage to life and property. Approximately 71 lakh persons in 2302 villages were affected by the loss of property, crop and other damages. 7932 persons lost their lives. In Krishna district alone 6706 persons died. The loss of cattle was 2,32,046 and 45,530 other livestock. There was substantial damage to houses—86,650 huts were completely washed away. Approximately 2,16,000 persons have either lost their houses or have suffered other loss. There has been widespread damage to crops worth of Rs. 450 crores and substantial damage to public buildings such as roads, electrical installations, highways, railways, telephonic installations, irrigation canals, etc.

To overcome the effects of 1977 cyclone, 172 relief camps were opened in Krishna, Guntur, East Godavari and Prakasam districts. About 2 lakh persons were provided shelter immediately. More than 25,000 quintals of rice and food packets were distributed. Medical special staff was sent to the affected areas to assist the Collectors to take measures against out-break of epidemics. Financial assistance was given for house repairs and the Government sanctioned remission of land revenue in the most affected areas of Machilipatnam, Repalle, Bapatla and Divi taluks.

Andhra Pradesh again experienced a major cyclone disaster in May 1979. Besides loss of life there was enormous damage to public and private properties in coastal districts. Prakasam, Nellore and Kurnool districts were the worst hit. The heavy rain under the influence of the cyclone, affected the districts of Guntur, Krishna, West and East Godavari on the coastline and Kurnool and Cuddapah districts of Rayalaseema and Mahaboobnagar district of Telengana. This was most unusual occurring in the month of May and again touching all the three regions in the state. Due to precautionary measures like providing information in time to the people of the state evacuating all the people to safer places mobilising and gearing up of administrative set up to take immediate steps, death toll and loss of property were greatly reduced though the intensity of the cyclone was double than that of the one which occurred in 1977.

Nearly Rs. 60 crores worth of crops were damaged in the State. The crops in Nellore district were heavily damaged followed by East Godavari, Guntur, Cuddapah and Prakasam districts. Due to intensive precautionary measures during precyclone period, the death toll was drastically kept under limit with just 750 human lives though the severity of it was very high compared to the earlier cyclones. About 3 lakh livestock were lost due to improper care taken during pre-cyclone and during cyclone periods. Out of the six cyclone affected districts, Prakasam lost heavily its livestock. Most of the deaths in this district were due to surging waters, which breached the tanks with extensive loss of livestock and damage to the property.

More than cyclone relief measures, 1979 cyclone reveals that the success of the excellent preparedness measures taken by the State could reduce the death toll greatly. Wholesale evacuation of the people living in low lying areas in coastal areas prevented not only the number of deaths and loss of property but also could reduce the risk of undertaking relief measures.

Another severe cyclone was experienced by the people of Andhra Pradesh in October, 1983. As Andhra Pradesh is called the rice bowl of India and more especially East and West Godavari districts, paddy crop was completely damaged. All the major rice producing districts such as East Godavari, West Godavari, Nizamabad and Karimnagar districts were severely affected besides Visakhapatnam, Khammam, Warrangal and other districts. Though the loss of human life was below 200, the cyclone could affect 50 lakh people in 6,322 villages of 14 districts. More than 2 lakh houses were completely damaged and nearly 1,50,000 houses were affected partially; 12,000 cattle and 31,000 livestock perished in the cyclone. Standing crops in 1,38,500 hectares were completely lost and crops in about 31,13,150 hectares were partially damaged in addition to dry crops in 3 lakh hectares of land. The loss of both public and private properties was estimated around Rs. 600 crores. The road length of more than 8,500 kilometres damaged completely. This was considered as one of the most severe cyclones in recent years as it could affect the economic structure of the State very badly.

Like that of 1979 cyclone, with utmost pre-cyclone preparation and protection, death toll and loss of property were minimised to the lowest possible level. Thanks to the immediate step taken by Telugu Desam Government and officials incharge in the districts, a major threat was averted with minimum loss by evacuating the people to safer places in almost all the coastal districts. Indian Army and Navy, Police and other officials could evacuate more than 50,000 persons in Visakhapatnam, East and West Godavari, Guntur, Krishna, Nizamabad, Karimnagar and Nalgonda districts to safer places. Hon'ble Chief Minister, Sri N.T. Rama Rao, has sanctioned immediately Rs. 30 crores towards relief without waiting for Central assistance.

After five months, in February 1984, the districts of Nellore and Chittor have again affected by a cyclone. The most unusual cyclone in the month of February could bring misery to farmers in more than 150 villages by destroying standing paddy crop and gardens worth of more than Rs. 20 crores and loss of property in the form of collapse of houses worth more than 50 crores. It is, however, the administration organised emergency relief and rehabilitation could set right the imbalance. The Cyclone of 1996 has the same impact on A.P. economy.

The cyclones which affected the people of Andhra Pradesh demonstrate that the people affected by cyclone disaster are not confined to the immediate geographic areas of destruction, death or injury. All persons who are related to, or who identify themselves with, persons and organisations in the stricken area are also affected. Thus the effectiveness of cyclone relief measures are usually national in scope. In each cyclone period, effective preparations were made to face cyclone with necessary arrangements by evacuating the population to safe places.

But there remains a problem of co-ordination and control in undertaking relief operations. Shortly after the cyclone impact, thousands of persons have to converge on the affected area and on first-aid stations, hospitals, relief centres and communication centres near the area. Along

with this movement of persons, incoming messages of anxious inquiry and offers of help from all parts of the country and even from foreign countries should come forward to set right communication facilities and to supply food, clothing, bedding and other materials. This action should continue for a few weeks following a cyclone. At times of cyclone generally there would be confusion due to lack of systematic procedures for maintaining a central strategic overview of the cyclone to apply controls and resources where they are most critically needed. Communication is often inadequate partly because of the destruction of communication facilities but more generally because of improper use of these facilities. With greater coordination and cooperation from the people and making them feel a great sense of urgency to help the victims, these effects of the cyclones would be greatly minimized in future.

# Bibliography


Allchin, B. and Allchin, R. *Civilization: India and the British of India Pakistan Before 500 B.C.* London, 1968.

All India Economic Conference. *Economic Life of Hyderabad.* Government Central Press, Hyderabad, 1937.

Andhra Pradesh, Government of, *Hand Book of Statistics,* Andhra Pradesh, 1993-94. p. 235.

Andhra Pradesh, Government of, *Economic and Statistical Bulletin.* Vol. XXXVII, No 11. July-December, 1992. Directorate of Economics and Statistics of Hyderabad. p. 10.

Andhra Pradesh, Government of, *Hand Book of Statistics-Andhra Pradesh,* 1993-94, Directorate of Economics and Statistics, Hyderabad, 1995. p. 7.

Andhra Pradesh, Government of, *Statistical Abstract of Andhra Pradesh,* 1992. Directorate of Economics and Statistics, Hyderabad, 1994. p. 111.

Andhra Pradesh, Government of, *Statistical Abstract of Andhra Pradesh,* 1992. Directorate of Economics and Statistics, Hyderabad, 1994. p. 111.

Agro-Economic Research Centre. *'Rice in Andhra Pradesh—A study on Inter-State Variations, Kharif,* 1978, *Part-1, Report",* Andhra University, Waltair, September, 1982. pp. 15-34 and 59-69.

Apinantara, Adul. *"Cooperation and Water Conflicts Among Water Users in North Eastern Thailand Tank Irrigation".* ADC, Workshop, Bangkok, Thailand, 1981 (Mimeo).

Appadorai, A. *Economic Research Centre in Southern India* (1000-1500) A.D. University of Madras, 1936.

Arputhraj, C. *"Problems and Prospects of Tank Irrigation in Tamil Nadu"*. Workshop on Modernisation of Tank Irrigation, Problems and Issues, Centre for Water Resources, Madras, 1982 (Mimeo)

Chambers, Robert, *"Men and Water: The Organisation and Operation"* in B.H. Farmer (ed). *Green Revolution*, Macmillan, London, 1977.

Cgrabheevyky, P. *Tank Irrigation and Agricultural Development.* Kanishka Publishing House, New Delhi, (India), 1992.

Coward, Sr.E. Walter, *Irrigation and Agricultural Development in Asia.* Cornell University Press, U.S.A, 1980.

Das Gupta, Kalyan Kumar, et. al. *"Problems of Management of Tank Irrigation", Impact of Tank Irrigation: A Case Study of Tumkur District.* Himalaya Publishing House, New Delhi, 1978.

Dean. D.J. and Fray. T.L. *"Discriminant Analysis of Loans for Cash Grain Farmers"*. Agricultural Economic Review, Vol. 36 (Annual) (April 1976).

Doherty, Victor S. *"A Cross Cultural Analysis of Tank Irrigation", Workshop on Modernisation of Tank Irrigation: Problems and Issues"*. Centre for Water Resources, Madras, India, 1982.

Department of Agricultural Engineering. *"Development and Optimisation in the use of irrigation Water under ex-zamin Tank in Ramanathapuram District"*, Preliminary Report, Government of Tamil Nadu, Madras, India, 1982.

Elumalai, G. *"Modernisation of Tank Irrigation Systems—Farmers' Views", Workshop on Modernisation of Tank Irrigation: Problems and Issues.* Centre for Water Resources, Madras, 1982 (Mimeo)

Evaluation and Applied Research Department, Government of Tamil Nadu. *"Evaluation of Minor Irrigation Schemes"*. Second Regional Workshop on Evaluation, Madras, India, 1979.

George, P.T., Namasivayam. D. and Ramachandraiah G. *"Application of Discriminant Function in the Farmers' Repayment Performance: A study in Chingleput District, Tamil Nadu"*, Journal of Rural Development, Vol. 3 No. 3 (May 1984).

George W. Snedecor and William G. Cochran, *Statistical Methods*, New Delhi. Oxford and IBH Publishing Co., (Sixth Edition), 1967.

Gupta, S.C. *Development Banking for Rural Development*, New Delhi, Deep and Deep Publications, 1987.

*Hand Book of Statistics: Anantapur District*, 1991-92 & 1992-93, Chief Planning Office, Anantapur, n.d.p. IV.

Harris, D.G. *Irrigation in India*, Oxford University Press, London, 1923.

Hayami, Y., Bennagem, E. and Barker, R. *"Price Incentive Versus Irrigation Investment to Achieve Food Self-Sufficiency in Philippines"*, American Journal of Agricultural Economics, Vol. 59, No. 4, 1977, pp. 717-721.

India, Government of. *Eighth Five-Year Plan, 1992-97*, Vol. II. 1992. p. 56.

India, Government of, *"Report on Minor Irrigation Works in the State of Madras."* Committee on Plan Projects, 1959.

India, Government of, *Census of India 1991-Final Population Totals*, Vol. II. Series 1, Paper 1, 1992, Ministry of Home Affairs, Government of India. New Delhi. p. 115.

India, Government of, Programme Evaluation Organisation. *A Study of the Problems of Minor Irrigation*, Planning Commission Publication. No. 140. 1961, pp. 7-103.

*Indian Agricultural Statistics*, for the years 1891-92 to 1947-48

Janakarajan, S. *"In Search of Tanks: Some Hidden Facts"*. Economic and Political Weekly, Vol. XXVII No 26. 1993. pp. A-53—A-60

Jayabalan, B.A. *"Modernisation of Tank Irrigation in Tamil Nadu"*, Workshop on Modernisation of Tank Irrigation Problems and Issues. Centre for Water Resources. Madras, 1980.

Jha, U.M., *Importance of Irrigation in an Agrarian Economy, Irrigation and Agricultural Development*. Deep and Deep, New Delhi, 1984.

Lakadawala, D.T., *"Growth, Empolyment and Poverty"*, Presidential Address, All India Labour Economics Conference, Tirupati, December, 1977.

Madras Institute of Development Studies. *"Tank Irrigation in Tamil Nadu—Some Macro and Micro Perspectives"*. (Report-Mimeo), 1983.

Mukundan, T.M., *"The ERY System of South India,"* PPST Bulletin, Madras September, 1988.

Namerta. *"Growth and Spatial Pattern of Market Towns in Rayalaseema Region of Andhra Pradesh,"* Unpublished M.Phil. Thesis submitted to Jawaharlal Nehru University (Centre for Development Studies), New Delhi, 1989.

Narayana Rao, J.S., *"Irrigation and Planning in Andhra Pradesh"*, Conference Papers of Andhra Pradesh Economic Association, Sixth Annual Conference, 23-24 January, 1988, pp. 145-150.

National Institute of Rural Development, *Rural Development Statistics,* 1994.

Narain, Dharan, and Roy B Shyamal, *"Impact of Irrigation on Labour Availability and Multiple Cropping"*, International Food Policy Research Institute, Research Report, 2-0, 1980, pp. 7-8 and 26.

*"Need for Restoration of Tanks."* The Hindu, August 8th, 1991, p. 3.

Narain, et al., *"An Approach to Study of Irrigation—A Case Study of Kanyakumari District,* Economic and Political Weekly, Vol. XVII. No. 39, September 25, 1982. p. 485.

Palaniswamy, K., *"Irrigation Tank Rehabilitation"*, The New Irrigation Era, Vol XIX, No. 3, 1981, pp. 36-37.

Palaniswamy, K. and Easter, K.W., *"The Tanks of South India: A Potential for Future Expansion in Irrigation"*, Economic Report, Department of Agriculture and Applied Economics, University of Minnisote, No. ER 83.4, 1983.

Rao, G.N. and Rajasekhar, D., *"Tank Irrigation in Andhra Pradesh: A Case Study"* Conference Papers of Andhra Pradesh Economic Association, Second Annual Conference, January, 1985, pp.57-70

Rao, V.M., *"Linking Irrigation with Development—Some Policy Issues"*. Economic and Political Weekly, Vol. XII, No. 24, 1978, pp. 993-97.

Rajasekhar, D., *Land Transfer and Family Partitioning,* Oxford and IBH Publishing Company, New, Delhi and Centre for Development Studies, Trivandrum, 1988.

Rao, N.V.N. and Ram Reddy, R., *"Minor Irrigation and Tribal Development—An Empirical Study"*, Kurukshetra, Vol. XXXIV, No. 2, November 1995, pp. 31-33.

Ramaswamy, V., et al., *"Damasi : A Concept of Equity and Productivity in Irrigation"*, Wamana, Vol. V, No. 3, July, 1985. pp. 1 and 15-22.

Ramanathan, *"Tank Irrigation in Tamil Nadu"*. Unpublished M. Phil. Thesis, Jawaharlal Nehru University, 1985, pp. 70-137.

Reddy, Narasimha, D., *"A Note on the Decline of Tank Irrigation"*, Paper presented at Lokayan Workshop on Tank Irrigation, Bangalore, July 1998.

Rao, V.M., *"Linking Irrigation with Development—Some Policy Issues"*, Economic and Political Weekly, Vol. XII, No. 24, July 17, 1978, pp. 993-997.

Rami Reddy.S., *"Factors Discriminating Defaulters from Non-defaulters in Primary Credit Cooperatives"*, Indian Cooperative Review, Vol. 14. No.1 (October 1976).

Satis, S. and Sundar, A., *People's Participation and Irrigation Management: Experiences, Issues and Options*, Commonwealth Publishers, New Delhi, 1990.

Sakthivadivel, R., et al., *"A Pilot Project Study of Modernisation of Tank Irrigation: Problems and Issues"*, Centre for Water Resources, Madras, 1982. (Mimeo).

Sen Gupta, Nirmal, *Managing Common Property: Irrigation in India and the Philippines*, Sage Publications, New Delhi, 1991.

Sen Gupta, Nirmal, *"Irrigation: Traditional Vs. Modern"*, Working Paper No. 55, Madras Institute of Development Studies, Madras, April 1985, p. 5 (Mimeo).

Sen Gupta, Nirmal, *Tank Irrigation in Gangetic Bihar*, A.N.S. Institute of Social Studies, Patna, Bihar, 1982.

Sivanappan, R.K., *"Water Management in Tank Irrigation in Tamil Nadu"*, Workshop on Modernisation of Tank Irrigation—Problems and Issues, Centre for Water Resources, Madras, 1982 (Mimeo).

Sivamohan, M.V.K., and Christopher A. Scott, (Ed), *India : Irrigation Management Partnerships,* Booklinks Corporation, Hyderabad, 1994.

SPSS Inc., SPSS/PC Release 1,1 update, the Author, U.S.A., 1984, pp. B. 202-204.

Sundar, A., and Rao, P.S., *"Farmers' Participation in Tank Irrigation in Karnataka"*. Workshop on Modernisation of Tank Irrigation: Problems and Issues, Centre for Water Resources, Madras, India,1982.

Srinivasan, T.M., *Irrigation and Water Supply: South India 200 B.C.–1600 A.D.,* New Era Publications, Madras. 1991.

Subbalakshmi, V., *"Incorporation of India and Indigenous Irrigation Institutions: The case of Dasabandam in Rayalaseema"*, Conference Papers of Andhra Pradesh Economic Association, Sixth Annual Conference, January, 1988, pp. 114-128.

Smith R. Baird, *Irrigation in Southern India,* Elder & Co., London. 18567. Ch 6.

Tubpin, Y., Easter, K.W., Welsch, D., *"Tank Irrigation in North-Eastern Thailand—The Returns and their Distribution"*, Economic Report, Department of Agricultural and Applied Economics, University of Minnesota, No Er, 82-6, 1982, p. 66.

Uma Shankari, *"Tanks: Major Problems in Minor Irrigation"*, Economic and Political Weekly, Vol.XXVI, No. 39, September 28, 1991, pp. A 115-A 125.

USAID, *"Thailand North-East Small Scale Irrigation"*, Washington, D.C., 1980.

Vasudeva Rao, D. *Rural Development Through Irrigation,* New Delhi, Ashish Publishing House, 1987.

Venkatram, B.R., *"Administrative Feasibility of Tank Irrigation Authority"*, Economics Programme, Consultancy Report, ICRISAT, Hyderabad, Andhra Pradesh, 1980, pp. 1-4.

Von Oppen, M. and Subba Rao, K.V., *"Tank Irrigation in Semi-Arid Tropical India, Part-I: Historical Development and Spatial*

*Distribution.*" ICRISAT,' Economics Programme Progress Report, 5, Andhra Pradesh, India, 1980. p. 1.

Von Oppen, M. and Subba Rao K.V., *"Tank Irrigation in Semi-arid Tropical India, Part I, II and III"*, Progress Report, ICRISAT, Hyderabad, 1980.

Von Oppen, M. and Subba Rao, K.V., *"History and Economics of Tank Irrigation in Semi-Arid Tropical India"*. Symposium on Rain water and Dry land Agriculture, Indian National Science Academy, New Delhi, 1982, pp. 89-93.

Von Oppen, M., *"Tank Irrigation in Southern India: Adopting a Traditional Technology to Modern Socio-economic Conditions"*, Proceedings of the Consultants' Workshop on State-of-the-Art and Management Alternatives for Optimising the Productivity of SAT Alfisoils and Related Soils, ICRISAT, Patacheru, Hyderabad, Andhra Pradesh, India. 1987, pp. 89-93.

Wijayaratna, C.M., *"Uneven Distribution of Water: Causes, Consequences and Implications of System Management"*, Workshop on Water Management, Agrarian Research and Training Institute, Colombo, Sri Lanka, 1982, (Mimeo).

# Index

N



O



P



R